TRIBOLOGY OF ELASTOMERS

TRIBOLOGY AND INTERFACE ENGINEERING SERIES

Vol. 24	Engineering Tribology (Stachowiak and Batchelor)
Vol. 25	Thin Films in Tribology (Dowson et al., Editors)
Vol. 26	Engine Tribology (Taylor, Editor)
Vol. 27	Dissipative Processes in Tribology (Dowson et al., Editors)
Vol. 28	Coatings Tribology – Properties, Techniques and Applications in Surface Engineering (Holmberg and Matthews)
Vol. 29	Friction Surface Phenomena (Shpenkov)
Vol. 30	Lubricants and Lubrication (Dowson et al., Editors)
Vol. 31	The Third Body Concept: Interpretation of Tribological Phenomena (Dowson et al., Editors)
Vol. 32	Elastohydrodynamics – '96: Fundamentals and Applications in Lubrication and Traction (Dowson et al., Editors)
Vol. 33	Hydrodynamic Lubrication – Bearings and Thrust Bearings (Frêne et al.)
Vol. 34	Tribology for Energy Conservation (Dowson et al., Editors)
Vol. 35	Molybdenum Disulphide Lubrication (Lansdown)
Vol. 36	Lubrication at the Frontier – The Role of the Interface and Surface Layers in the Thin Film and Boundary Regime (Dowson et al., Editors)
Vol. 37	Multilevel Methods in Lubrication (Venner and Lubrecht)
Vol. 38	Thinning Films and Tribological Interfaces (Dowson et al., Editors)
Vol. 39	Tribological Research: From Model Experiment to Industrial Problem (Dalmaz et al., Editors)
Vol. 40	Boundary and Mixed Lubrication: Science and Applications (Dowson et al., Editors)
Vol. 41	Tribological Research and Design for Engineering Systems (Dowson et al., Editors)
Vol. 42	Lubricated Wear – Science and Technology (Sethuramiah)
Vol. 43	Transient Processes in Tribology (Lubrecht, Editor)
Vol. 44	Experimental Methods in Tribology (Stachowiak and Batchelor)
Vol. 45	Tribochemistry of Lubricating Oils (Pawlak)
Vol. 46	An Intelligent System For Tribological Design In Engines (Zhang and Gui)

Aims & Scope

The Tribology Book Series is well established as a major and seminal archival source for definitive books on the subject of classical tribology. The scope of the Series has been widened to include other facets of the now-recognised and expanding topic of Interface Engineering.

The expanded content will now include:
• colloid and multiphase systems; • rheology; • colloids; • tribology and erosion; • processing systems; • machining; • interfaces and adhesion; as well as the classical tribology content which will continue to include • friction; contact damage; • lubrication; and • wear at all length scales.

TRIBOLOGY AND INTERFACE ENGINEERING SERIES, 47
EDITOR: B.J. BRISCOE

TRIBOLOGY OF ELASTOMERS

SI-WEI ZHANG
University of Petroleum
Beijing, People's Republic of China

ELSEVIER

2004

Amsterdam – Boston – Heidelberg – London – New York – Oxford
Paris – San Diego – San Francisco – Singapore – Sydney – Tokyo

ELSEVIER B.V.
Sara Burgerhartstraat 25
P.O. Box 211, 1000 AE
Amsterdam, The Netherlands

ELSEVIER Inc.
525 B Street, Suite 1900
San Diego, CA 92101-4495
USA

ELSEVIER Ltd
The Boulevard, Langford Lane
Kidlington, Oxford OX5 1GB
UK

ELSEVIER Ltd
84 Theobalds Road
London WC1X 8RR
UK

First edition 2004

Library of Congress Cataloging in Publication Data
A catalog record is available from the Library of Congress.

British Library Cataloguing in Publication Data
A catalogue record is available from the British Library.

ISBN: 0-444-51318-3
ISSN: 1572-3364 (series)

∞ The paper used in this publication meets the requirements of ANSI/NISO Z39.48-1992 (Permanence of Paper).
Printed in The Netherlands.

PREFACE

When considering the achievement of national economic growth and the improvement of the quality of people's lives, rubber figures as one of the most important raw materials. At present, the production of a wide variety of tyres destined for the carriage of both people and goods, constitutes its main application. In addition to being a material of high elasticity, rubber, however, possesses many unique properties which metals and other polymeric materials lack. This, in turn, has led the ever more extensive application thereof to a great variety of friction components in machinery and to the fact that it currently occupies a prominent position amongst the contemporary engineering materials. Questions relating to the improvement of the tribological performance and the enhancement of the service life of a wide range of rubber products, have accordingly been engendering progressively increasing levers of interest and attention.

Despite of the high practical relevance of considerations such as the friction, wear and lubrication of elastomers, a paltry few monographs are available thereon. Moreover, these were published more than twenty years ago and offer but a partial consideration of tribology-related topics. There have, however, since then been remarkable advances in the field of tribology. Given the latter, a new and comprehensive work on rubber tribology has obviously long been in the waiting.

Based on original rubber wear studies over the past twenty years, the year 2000 saw the publication in China, of a work entitled "Wear of Elastomers". In order to provide a more recent and complete understanding of the friction, wear and lubrication of elastomers, the present work, "Tribology of Elastomers", is now offered. As for contents, it constitutes both a widening of the scope of and a supplement to the former. It covers the three main facets of rubber tribology(i.e. friction, wear and lubrication). In the main, it contains the original research results attained by the author, while cognizance is also taken of some of the related achievements of other scientists and available basic literature in this field. This work fills a gap in the currently available literature on tribology and reflects the latest developments in that field.

In this work, the information on the friction, wear and lubrication of elastomers is presented systematically and in logical order. However, inasmuch as the knowledge of rubber wear lags behind that of other aspects of tribology of elastomers, this work essentially focuses that aspect, particularly as regards the mechanisms and theories relating thereto. Adequate understanding of these aspects constitutes an obvious prerequisite to the development of rubber formulation design principle and the methods for testing and estimating the wear resistance of rubber, and to go a step further towards increasing the working life of rubber products.

Chapter 1 provides some background to tribology of elastomers. In Chapter 2, two types of typical friction, i.e. point-contact and line-contact friction are introduced, as well as two kinds of special friction, namely run-in friction and friction on ice and snow, while Chapter 3 addresses the basics of wear. Given the prevalence of rubber in practice, Chapter 4 deals with three types of rubber abrasion, with the theories relating to rubber abrasion and computerized analytical methods for worn surface and wear debris, which are utilized for quantitative abrasion process analysis, being presented in Chapter 5. The remaining three types of rubber wear, namely erosion, fatigue wear and frictional wear, are discussed in Chapters 6 and 7. The particularly important surfacial mechanochemical effect of abrasive erosion is described in Chapter 8 with reference to four kinds of elastomer, the said effect being highly conducive to a more enhanced understanding of the wear mechanisms of abrasive erosion. The wear of metal by rubber, being a special aspect which has long been ignored, is briefly introduced in Chapter 9, while the lubrication of rubber seals, given the wide-spread use of seals in hydraulic and other fluid-oriented machinery, constitutes the focus of Chapter 10.

The book is mainly intended for researchers and engineers who are interested in rubber tribology and its application. It could also well serve as a text or supplementary text for teachers and students at tertiary institutions who wish to familiarize themselves with tribology of elastomers as an interdisciplinary field of research and activity.

I sincerely trust that the present monograph will help to create and stimulate a new international dialogue on rubber tribology and related fields.

Si-wei Zhang

Beijing
April, 2004

ACKNOWLEDGMENT

I would like to express my deepest gratitude to Professor A.N. Gent, from the University of Akron, and Professor K.C. Ludema, from the University of Michigan, in the United States, and also to Professor B.J. Briscoe, of the Imperial College of Science, Technology and Medicine in the United Kingdom, for their full support and kind assistance. It was on account of their great assistance that I could and did fulfill my research work abroad in the 1980's.

Special thanks are due to Professor B.J. Briscoe for his encouragement and suggestion to publish this book outside my own country. I would also like to thank Professor D. Dowson and Mr. D. Eastbury for giving me the opportunity to include this book in the Tribology and Interface Engineering Book Series.

Thanks are extended to past coworkers and students for their participation in developing the topics and concepts presented in this book. Sincere thanks are also due to Professor X.D. Peng from the University of Petroleum in China for contributing the draft of a section on tyre traction force on ice and snow.

I am very grateful to the many authors and publishers whose kind permission enabled me to publish some of the illustrations and tables in this book. Acknowledgements are presented in the figure and tables captions and in the list of references.

Finally, I would like to thank Miss Dong Shanying and Miss Ma Hongyu for their assistance in typing the preliminary draft. I wish to extend special acknowledgment to Mr. Zhu Runmin for his services, not only in typing the preliminary draft but producing the camera-ready manuscript as well.

Si-wei Zhang

Beijing
April, 2004

CONTENTS

Chapter 1 INTRODUCTION 1

1.1 Concept of Elastomers 1
1.2 Definition of Tribology of Elastomers 2
1.3 Significance of Tribology of Elastomers 2
1.4 An Essential Approach for Studying in Tribological Problems - Systems Analysis 3
1.5 Basic Features of Elastomers 5

Chapter 2 FRICTION 7

2.1 Definition and Essential Characters 7
2.2 Run-in Friction 8
2.3 Rubber Friction by a Point Contact 14
2.4 Rubber Friction by a Line Contact 24
2.5 Tires Traction (Friction) Forces on Ice and Snow 28

Chapter 3 AN INTRODUCTION TO WEAR 33

3.1 Definition and Essential Characters 33
3.2 Classification 37

Chapter 4 ABRASION 39

4.1 Dry Abrasion 39
4.2 Oily Abrasion 66
4.3 Wet Abrasion 69

Chapter 5 THEORY OF RUBBER ABRASION ... 85

5.1 Fatigue-Fracture Theory ... 85
5.2 Energy Theory ... 100
5.3 Fractal Analysis and Computer Simulation of Worn Surfaces ... 110
5.4 Computer Image Analysis of Wear Debris ... 126

Chapter 6 EROSION ... 135

6.1 Mechanisms of Abrasive Erosion ... 135
6.2 A Brief Introduction to Theory of Metal Erosion ... 150
6.3 Theory of Abrasive Erosion ... 156

Chapter 7 FATIGUE WEAR AND FRICTIONAL WEAR ... 177

7.1 Fatigue Wear ... 177
7.2 Frictional Wear ... 180

Chapter 8 SURFACIAL MECHANOCHEMICAL EFFECTS OF ABRASIVE EROSION ... 185

8.1 Introduction ... 185
8.2 Surfacial Mechanochemical Effects of Nature Rubber ... 187
8.3 Surfacial Mechanochemical Effects of Nitrile Rubber ... 199
8.4 Surfacial Mechanochemical Effects of Fluororubber ... 206
8.5 Surfacial Mechanochemical Effects of Polyurethane ... 213

Chapter 9 WEAR OF METAL BY RUBBER ... 227

9.1 Literature Survey ... 227
9.2 Wear of Metal by Rubber under Boundary Lubrication Condition ... 236

Chapter 10 LUBRICATION OF RUBBER SEALS ... 247

10.1 Hydrodynamic Lubrication of O-ring Seals for Reciprocating Motion ... 247
10.2 Visco-Elastohydrodynamic (VEHD) Lubrication in Rotary Lip Seals ... 249
10.3 Micro- Elastohydrodynamic (MEHD) Lubrication in Rotary Lip Seals ... 252

References ... 259

Subject Index ... 267

Chapter 1

INTRODUCTION

1.1 Concept of Elastomers

"Elastomer" is usually used as a general term for the group of polymers with some common characters, such as high elasticity, viscoelasticity and glass transition temperature far below room temperature. In general, rubber might be called elastomer since the high elasticity is its most outstanding feature.

Nevertheless, strictly speaking, we cannot regard the concept of elastomer as the same as that of rubber in the narrow sense, because the former refers to all the polymeric materials with high elasticity including rubber. However, by usage, it is generally admitted in a broad sense that the term "rubber" refers to elastomer. In this book, the two words are interchangeable since the word "rubber" is referred to the broad concept of "rubber" more often than not.

Natural rubber (latex) was the first kind of elastomer utilized in engineering practice and vulcanizate followed. The latter is a kind of thermosetting elastomers, which cannot be melted any more after solidification once only. During the Second World War, in imitation of natural rubber, a series of synthetic rubbers with more wide-ranging performance were developed for industrial application, which are also a kind of thermosetting elastomers. Although both polyurethane elastomers developed in the 1950's and the other ones produced later on have high elasticity too, their other properties are different from that of rubber. Moreover, some of them have much the same structure as thermoplastic and thermosetting plastics. Afterwards, the development of thermoplastic elastomers marked that the processing industry of elastomers extended great progress, because this kind of elastomers can be processed utilizing completely various techniques processed thermoplastic plastics. These techniques are repeated melting technology as well as molding and extruding techniques including vacuum forming, blow molding and high speed injection molding and so on. Consequently, the production cost was greatly reduced.

1.2 Definition of Tribology of Elastomers

It is well known that the original definition of tribology is "the science and technology of interacting surfaces in relative motion and the practices related thereto" [1]. Therefore, tribology of elastomers as a branch of tribology could be defined as "the science and technology for investigating the regularities of the emergence, change and developing of various tribological phenomena in rubber and rubber-like materials and their tribological applications". Certainly, these tribological phenomena are brought about by a combination of interaction between the interacting surfaces in relative motion and the environment, including not only mechanical and physical, but chemical, thermochemical, mechanochemical and tribo-chemical as well.

In view of the engineering applications, elastomer tribology might also be considered as a science and technology for studying the tribological behaviors of elastomeric materials and their engineering practices.

Tribology of elastomers is a growing interdisciplinary area, which of possible related interests are rubber and rubber-like materials and its composites, mechanochemistry, surface physics, surface chemistry, fracture mechanics and polymer physics, etc.

1.3 Significance of Tribology of Elastomers

Rubber is an indispensable and important raw material to the development of industry and human life. Since the 1950's, the consumption of rubber has had a close and stable correlation with Gross National Production, which reflects the development of the national economy to a certain extent. The conventional use of rubber is mainly manufacturing tires employed by various vehicles, while the automobile industry is one of the pillar industries of the national economy in many countries.

As a kind of high elasticity materials, rubber has some favorable properties (such as wearablity, oil-resistance, etc.), which metal and other polymeric materials lack. Therefore, it is widely used not only in automobile industry, but also in other industries. From the 1970's on, the application of rubber to vehicle tires and various frictional components, such as sealing rings, soft limed journal bearings, water lubricated bearings and so on, retained a continuously increasing trend all the time. Even if the wear-resistance and service life of rubber are not increased so much, the considerable economic and social benefits can be brought out in many respects, including the conservation of energy, materials and lubricants.

In addition, since the tribology of elastomers is an important growing interdisciplinary area, the developing of rubber and its products as well as tribology will certainly simulate its growing and increase demand for information in this field. Therefore, a clear and complete understanding of rubber tribology is of practical value and becoming increasingly important.

1.4 An Essential Approach for Studying in Tribological Problems–Systems Analysis

In engineering practices, the tribological problems are usually quite complicated. Moreover, the analysis of any tribological phenomenon cannot separate itself from the related realistic tribo-system. Therefore, systems analysis must be considered as an essential approach to the tribological problems.

As early as the end of the 1970's, Czichos expounded systematically the principle and methodology of systems analysis applied to tribology[2]. However, the present author held that on the basis of the fundamentals of systems engineering, there is a need to introduce some more concepts of systems engineering and set theory in order to describe and evaluate a tribo-system more completely and precisely [3, 4].

A tribo-system is a set of elements interconnected by a certain kind of structure to constitute an organic whole for the sake of performing a certain function. These elements may be parts, components or subsystems, but must at least contain a couple of frictional pairs with interacting surfaces in relative motion.

Any system can be divided into some subsystems in different ranks according to its different function. Therefore, a tribo-system F composed of n subsystems f_i may be a subsystem of a larger system L. Then

$$f_i \subset F \subset L \qquad (i = 1,2,3,\cdots,n) \tag{1.1}$$

In the view of systems engineering, only by involving all of the three essentials of a tribo-system, namely structure, function and objective function or evaluation index, can we describe this system completely.

The structure of a system is a description of the interior characteristics of the system. It is represented by a set of its elements and their relevant properties as well as the relations between an element and the other ones, and the relations between each element and the system, namely

$$S = C \cup R = \{x | x \in C \vee x \in R\} \tag{1.2}$$

Where,

$$C = A \times P = \{(a, p) | a \in A, p \in P\} \tag{1.3}$$

$$A = \{a_1, a_2, \cdots, a_n\} \tag{1.4}$$

$$P = \{p_1(a_i), p_2(a_i), \cdots, p_m(a_i)\} \tag{1.5}$$

$$R = \{r_1(a_1, \cdots, a_j), r_2(a_1, \cdots, a_j), \cdots, r_k(a_1, \cdots, a_j), R_1(a_i), R_2(a_i), \cdots, R_k(a_i)\} \tag{1.6}$$

Where, x, a_i, p_i, r_i and R_i are elements, $i = 1, 2, \cdots, n$; $2 \le j \le n$.

The relation between sets *A* and *P* can be summed up as several types shown in Figure 1.1.

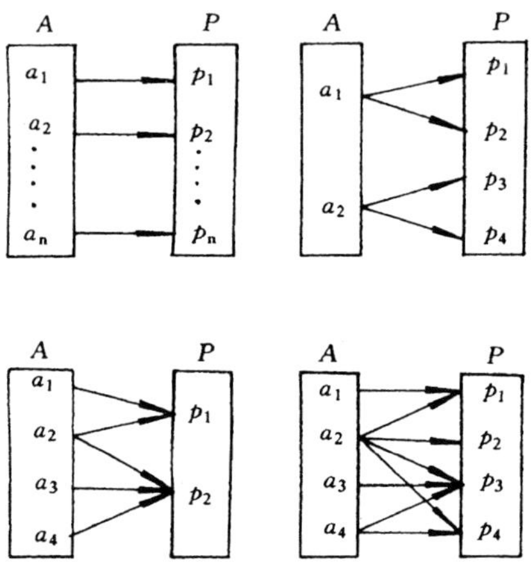

Fig. 1.1. Correlatograph of the set of elements *A* with the set of the relevant properties of the elements *P*

The function of a system is a description of external characteristics of the system. It shows the transformation of inputs *X* into outputs *Y*, and can be expressed as

$$\{X\} \xrightarrow{I} \{Y\} \tag{1.7}$$

Where, ***I*** is the input-output transferring function.

As for the objective function or evaluation index of a system, it is used to evaluate the effectiveness of the system.

The procedure of systems analysis can be generally expressed in the following block diagram (Figure 1.2).

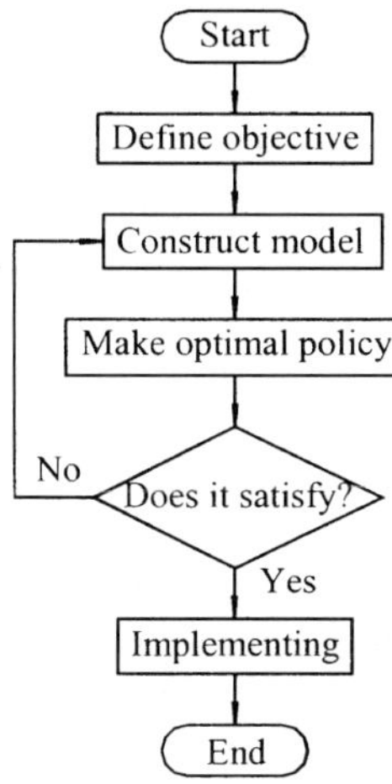

Fig. 1.2. Block diagram of systems analysis

1.5 Basic Features of Elastomers

Rubber is a macromolecular compound consisted of many macromolecules. Each macromolecule is a very long macromolecular chain that is formed from a number of chemical structural units bound by covalent bonds. For example, a molecular chain of natural rubber consists of about 1000 to 5000 isoprene chemical structural units.

The structure of a single macromolecular chain has three basic forms (Figure 1.3), namely linear (straight-chain) macromolecule, branched (branched-chain) macromolecule and crosslinked macromolecule.

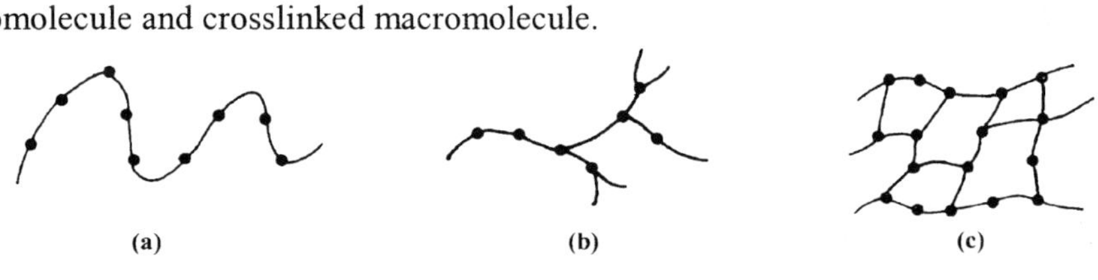

Fig. 1.3. Basic forms of macromolecular structure
(a) linear macromolecule; (b) branched macromolecule;
(c) crosslinked macromolecule

The molecular chains of linear macromolecule are quite easy to move each other. Therefore, the linear macromolecule can be softened with heat, and hardened with cooling. This characteristic is called thermoplasticity. As for the crosslinked macromolecule, the relative motion among its molecular chains is greatly restricted, thus it cannot flow or melt easily being heated. This behavior is called thermoset.

Compared with metals, rubber generally possesses the following features:

(1) Elastic deformation is very large, while elastic modulus is extremely small. The elastic deformation of rubber can be up to 1000 %, and that of most polymeric materials is only 1 % or so. As for the ordinary metals, the elastic deformation is smaller than 1 %. The elastic modulus of rubber is only about 10^6 times less than that of metals, and it is increased in direct ratio with temperature, yet the elastic modulus of metals is opposite to that of rubber.

(2) The Possion's ratio of rubber (0.49) is larger than that of ordinary metals, near to that of liquid (0.5). Therefore, during deformation, the volume of the rubber is almost unchanged, while the metals are different from this.

(3) Elastic deformation of uncrosslinked rubber presents distinct time-dependence, namely relaxation properties, which the metals do not have.

(4) The thermal effect is more evident during deformation of the rubber. It expresses as that quickly stretching of the rubber results in the removal of heat and restoring itself causes the absorption of heat, yet the metals are quite the opposite.

Table 1.1 gives the main mechanical properties of several kinds of rubbers being in common use [5].

Table 1.1. Main mechanical properties of several kinds of rubbers

Kinds of rubbers	Shore -hardness	Tensile strength (MPa)	Maximum service temperature (℃)	Glass transition temperature (℃)
Isobutylene- isoprene rubber	30-100	21	149	-75
Chloroprene rubber	40-95	21	116	-49
Nitrile rubber	20-90	21	149	-22
Silicone rubber	10-85	7	316	-120
Polyurethane	10-100	34	116	-
Natural rubber	30-100	25	70	-70
Ethylene propylene rubber	30-90	10	125	-58
Polybutadiene	35-90	10	70	-85

Chapter 2

FRICTION

Although the friction of elastomers has been studied extensively in basic mechanisms and practical applications, only a few literatures are available on some special subjects, such as run-in friction as well as rubber friction by a point contact and by a line contact. Therefore, these contents will be presented in this chapter, which are mainly based on the author's research achievements. In addition, an important and interesting subject, rubber friction on ice and snow, is discussed by generalizing the basic references.

2.1 Definition and Essential Characters

Friction is a very complicated phenomenon. Since Leonado da Vinci studied the friction of solids late in the fifteenth century, a universally accepted theory to clarify exactly the friction phenomena has not yet been developed up to now. There is not even a generally recognized understanding of the definition of friction. For instance, in the "Encyclopedia of Tribology", the term "friction" is defined as: the resisting force tangential to the common boundary between two bodies when, under the action of an external force, one body moves or tends to move relative to the surface of the other [6]. However, Kragelsky and Alisin considered that external friction of solids is a complex phenomenon depending on various processes that occur in the real areas of contact and in the surface layers during relative tangential movement of the bodies [7].

If we see the essence through the phenomena of friction, the friction is in essence an energy-transforming effect, not force. For example, when a train is slowing down, the friction occurring in the braking process is mainly an effect being transformed from mechanical energy into heat. Therefore, the definition of friction might be stated as follows: friction is an energy-transforming effect generated in the interface between contact objects or a body in contact with another substance in relative motion under the action of an external force. It is certainly resulted from the action of tangential resistance, namely, friction force, occurring in the interface. Thus it can be seen that friction and frictional force are two different concepts, and these two terms have both close relation and essential distinction. Hence, friction should not be confused with friction force.

2.2 Run-in Friction

Run-in friction is an unsteady state friction. It refers to such a friction state that the frictional coefficient varies as the sliding time (sliding distance) during the beginnings of relative motion. As early as the beginning of 1940's, Roth, Driscoll and Holt already examined the run-in friction for rubber sliding on glass tracks and pointed out the importance of study in this friction state [8]. However, this area was neglected for a long time. Not until the early 1980's did the problem of run-in friction of rubber come to the attention of scientists again. Cooper and Ellis [9] investigated run-in friction of carbon black reinforced rubber sliding against perspex and achieved much the same conclusions as Roth et al did, namely, the frictional coefficient of run-in friction gradually increases with sliding distance (sliding time) until a constant limiting value is reached. Later on, Ellis [10] studied the run-in friction on condition that the steel, glass and perspex spherical indenters were pulled across a rubber truck respectively and proposed an empirical equation for the corresponding coefficient of friction, but he did not clarify the physical process of this kind of rubber friction. However, for viscoelastic rubber, a deep understanding of the mechanism of run-in friction is necessary to grasp the nature of rubber friction comprehensively.

2.2.1 Characteristics of Run-in Friction

Experimental study in the basic characteristics of run-in friction between silicone rubber and glass was conducted by using a friction- testing machine as shown in Figure 2.1 [11]. The glass disc was rotated against a stationary silicone rubber hemisphere. For the sake of reducing the influence of friction heating, the rotating speed was controlled within the range of 6.8×10^{-3}~8.1×10^{-2} m/s. The normal load applied on a rubber sample was ranged over 1.18~4.9 N. Experiments were carried out at room temperature, 25℃~29℃.

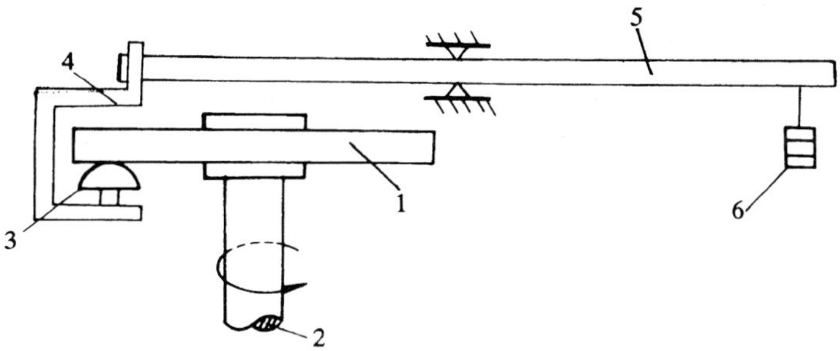

Fig. 2.1. Sketch of friction testing machine
1-glass disc; 2-spindle; 3-rubber sample; 4-sample support; 5-bar; 6-dead weights

In order to investigate the effect of the viscoelastic properties of rubber on the run-in friction, three different test procedures were used with variable stationary contact time between the rubber and the glass disc before generating relative motion.

1. Stationary loading

The glass disc is set into rotation once the rubber sample is pressed against its surface with a preset normal load. The stationary contact time is generally less than 3s.

2. Moving loading

The rubber sample is pressed against the surface of the glass disc with a preset normal load after the glass disc is set into rotation at a desired stable speed. Obviously, the stationary contact time in this procedure is equal to zero.

3. Lasting loading

The glass disc is set into rotation after the contact between the rubber sample and the glass disc is established with a preset normal load and lasts a desired time (including three situations of 1 min, 1 h and 15 h). Obviously, the stationary contact time in this procedure is longer.

The three test procedures above represent three initial contact states.

The basic characteristics of the run-in friction generated between silicone rubber and glass can be observed from the experiment results (Figures 2.2 and 2.3). It shows that the frictional coefficient is increased with the sliding time (run-in time) t or sliding distance until a steady maximum value (namely steady frictional coefficient) is attained in the initial period of movement.

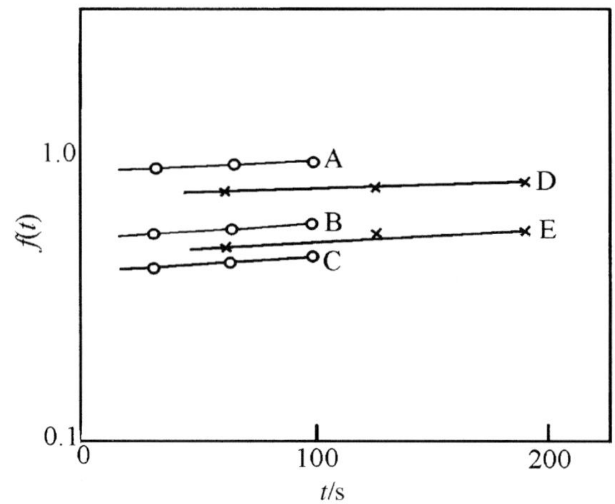

Fig.2.2. Frictional coefficient plotted against run-in time under the condition of different velocities υ and different normal loads N

A: $\upsilon=1.8\times10^{-2}$ m/s, N=1.18 N; B: $\upsilon=1.8\times10^{-2}$ m/s, N=2.49 N;
C: $\upsilon=1.8\times10^{-2}$ m/s, N=3.63 N; D: $\upsilon=9.4\times10^{-3}$ m/s, N=1.18 N;
E: $\upsilon=9.4\times10^{-3}$ m/s, N=2.49 N.

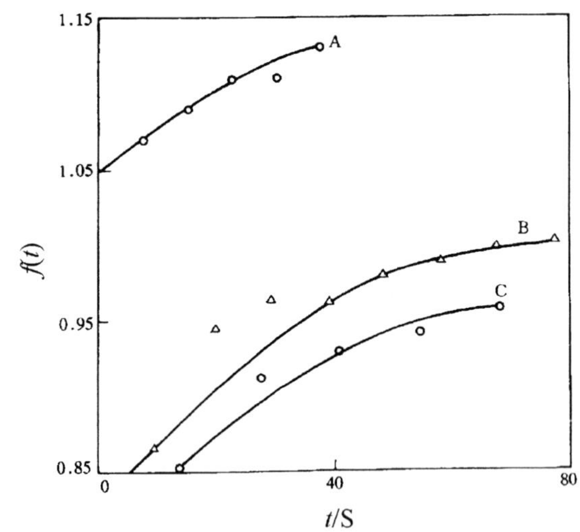

Fig.2.3. Frictional coefficient plotted against run-in time under the condition of different velocities (N=1.18 N)

A: $\upsilon=8.1\times10^{-2}$ m/s; B: $\upsilon=6.2\times10^{-2}$ m/s; C: $\upsilon=4.4\times10^{-2}$ m/s

The above results are coincided with that obtained by Ellis [10] even though being used different experimental device, samples, procedure and method. The basic characteristics of run-in friction of rubber do not change with the testing parameters such as N, υ, etc. Moreover, just like the steady friction, at low normal load, the frictional coefficient is decreased with an increase in normal load (Figure 2.2) and increased with the increment of sliding velocity (Figure 2.3).

On condition that $\upsilon \leq 1.8\times10^{-2}$ m/s, on the basis of Fig. 2.2, an expression of unsteady friction coefficient can be given to a good approximation as:

$$f(t) = a\exp(bt) \tag{2.1}$$

Where, a, b are constants between 0 and 1. Furthermore, a is increased with an increase in sliding velocity, and with a decrease in normal load. However, b is increased with the increment of both sliding velocity and normal load.

If only $\upsilon > 1.8\times10^{-2}$ m/s, the above equation is invalid. It would be accounted for the heat effect on the friction surfaces.

Based on Figure 2.2, another explicit correlation for unsteady frictional coefficient is deduced (Figure 2.4):

$$f(t) = \alpha N^{-\beta} \tag{2.2}$$

Where α and β are constants which vary with sliding velocity and α=1, β=0.7 under the condition of $\upsilon = 1.8\times10^{-2}$ m/s.

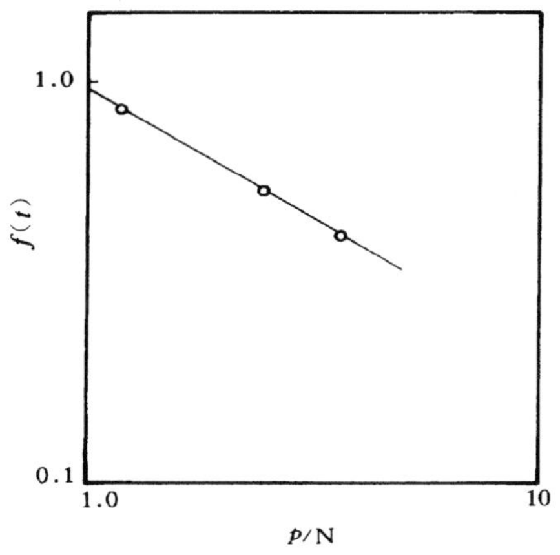

Fig. 2.4. Effect of normal load on frictional coefficient ($\upsilon = 1.8\times10^{-2}$ m/s)

It is merited attention that equation (2.2) is similar in form to the empirical relation of frictional coefficient in steady state at low load for polymers, as proposed previously by Tabor [12], except that the exponent lies between 0 and 1/3 in Tabor's equation. This fact confirms further that the run-in friction of rubber posses same characteristics as that of steady friction besides the time-dependent behavior.

2.2.2 Effect of Initial Contact State on Run-in Friction

As shown in the experimental results, different initial contact-states did not change the basic characteristics of the run-in friction of rubber, i.e., the unsteady frictional coefficient is still linear with the sliding time (sliding distance) (Figure 2.5).

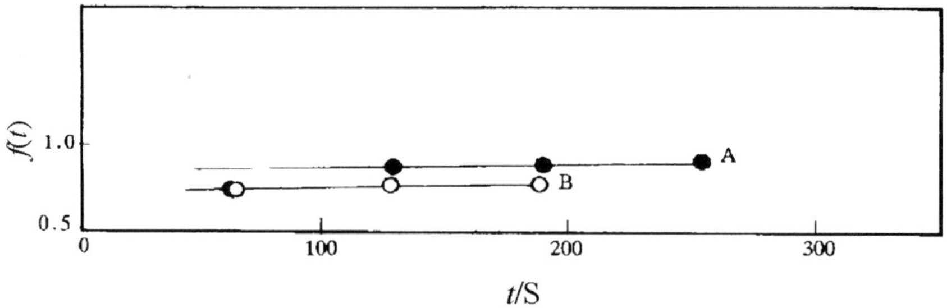

Fig. 2.5. Effect of test procedure on run-in friction of rubber ($N=1.18$ N, $\upsilon=0.94\times10^{-2}$ m/s)
A- second procedure, B- first procedure

However, if we make a comparison of the steady frictional coefficient f_m and run-in time t_s reached steady friction, which were obtained by using three different test procedures respectively under otherwise identical conditions (See Table 2.1), it can be found that the values of f_m and t_s reached by the second test procedure are the largest, and that the values reached by the third test procedure are the smallest. As seen from Figure 2.5, the unsteady frictional coefficient obtained by the second test procedure is also larger than that by the first test procedure. It is sufficient to show that the effect of the initial contact state on run-in friction of rubber cannot be ignored.

Table 2.1. Effect of test procedure on run-in friction ($N=1.18$ N, $\upsilon=0.94\times10^{-2}$ m/s)

Test procedure	f_m	t_s, s
1	0.76	190
2	0.88	255
3	0.69	128

The above experimental results probably give an expression to the viscoelastic effect of rubber, namely, the loaded contact between the rubber hemisphere and glass disc generates viscoelastic flow at the asperities of the friction surface of rubber to different extents and enhances the conformal effect to be mutually conformed between the rubber and its counterface. The longer the dwell time in keeping the initial contact state, the better the conformal effect is. As a result, the shorter the run-in time, the smaller the coefficient of kinetic friction for reaching steady friction is.

2.2.3 Hypothesis on the Physical Processes of Run-in Friction

The interaction of the rubber, as a kind of viscoelastic materials, with a rigid surface on a sliding contact interface is essentially different from that of the metals. Many scholars [13] consider this interaction to contain adhesion and hysteresis (deformation). Therefore, it could be deduced that the frictional force resulted from the interfacial interaction is mainly dependent on the intrinsic characteristics of frictional pairs as follows:

(1) Consistence of coupled materials;

(2) Contact behavior of interface. It depends not only on the morphology of solids, but the elasticity, viscoelasticity and surface hardness of rubber as well.

If a polymeric sphere slides over a relatively clean and smooth hard surface, most frictional forces are resulted from the interfacial adhesion [14]. Moreover, run-in friction is an unsteady friction, which means that the energy of interface is also constantly changing, even though the operating and environmental conditions are kept constant. As for why and how this energy changes, it is still obscure. For this reason, based on the theories of elastic physics and adhesion of rubber, the following hypothesis presented is intended to make a theoretical explanation for the physical processes of run-in friction of the rubber.

The run-in friction of rubber could be viewed as a process of this kind (Figure 2.6), namely the soft rubber sample conforms gradually with the rigid counterface (glass disc) along with the increase in sliding time, and this is characterized by a time-dependent adhering or shearing process. It is assumed that the surface of the glass disc is very smooth and clean. Moreover, it could be considered as a hard surface on which a number of spherical micro-asperities of uniform geometry are evenly distributed (Figure 2.6). On the assumption that the deformation of each asperity on the rubber surface is independent of each other, thus, under the action of normal load, any point on the surface of stationary rubber sphere contacted with the glass disc will engender sine-shaped alternate stress during relative movement. Provided the normal load is unchangeable, the alternate stress of every point is also kept constant.

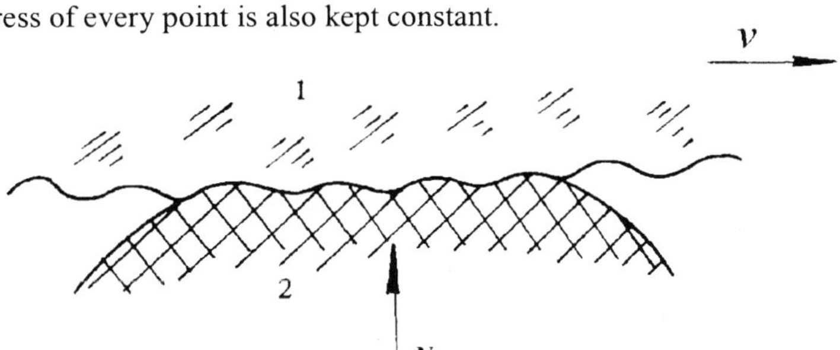

Fig. 2.6. Topography of an ideal contact surface
1-glass disc; 2-rubber sample

Under the condition that the frictional power is not large enough, the wear of the rubber surface is so small as to be neglected. In this case, the first stage of the run-in frictional physical process of rubber might be only viewed as a process of elastic and viscoelastic deformation. In regard to the deformation process, the rubber surface produces elastic deformation firstly and then generates time-dependent viscoelastic flow (deformation) under the action of normal load. Based on the above assumption, it could be regarded that under the condition that both normal load and sliding velocity are unchangeable, the deformation processes at each point on the contact surface of rubber are identical, and the elastic deformation is also fully equal, but the viscoelastic deformation is increased with time. As a result of the above deformation, the micro contact area of each micro-asperity is enlarged, thus increasing the total area of the true-contact surface on the apparent contact surface more and more. It should be pointed out that in this situation, the elastic deformation of a certain point on the contact surface of the rubber is emerged and disappeared periodically along with time, but the viscoelastic deformation is increased with sliding time until reaching a stable value finally.

The second stage of the run-in friction physical process is a periodical generating and shearing (damage) process of the adherence between the asperities of the rubber surface and the counterface. On the basis of the phenomenological theory of rubber adhesion [15], the adhesive friction in the molecular level mainly arises from the weak van der Waals' force under the condition of without thermal effect of friction, if the static electric force and hydrogen bond force are ignored. On the assumption that the adhesive work of each molecule is approximately equal to van der Waals' potential [16], as viewed from micro, the total adhesive work or adhesive frictional force is increased in the total true contact area on the frictional surface and reached a stable value to the last, namely frictional force of steady friction.

In addition, the height of each asperity on the rubber surface will decrease correspondingly along with the increase in the contact area of the asperity. As a result, the number of asperities on the rubber surface in contact with the glass disc increases in some degree, which also enhances the total true contact area further.

2.3 Rubber Friction by a Point Contact

2.3.1 Theoretical Analysis of Friction Process

As stated above, the interaction between rubber and a rigid surface on the sliding contact interface includes adhesion and hysteresis (deformation). This section will only discuss a simple friction process of the tribo-systems. In this process, the rubber is in the state of a point contact with a single stiff asperity on the counterface and no continuous action of the other asperities on the counterface against the rubber surface. Therefore, the effect of the rubber hysteresis on friction could be neglected. However, since the sharp asperities can insert into the rubber surface to cause material losses [17,

18], it can be held that the friction force occurred on the friction surface of rubber is mainly comprised of two parts. One is the micromolecular force generated on the interface between the rubber and its counterpart, and another is the macro mechanical force acted on the rubber surface by the counterpart of the rubber. The corresponding physical process includes the adhesion process of rubber with its counterface and the rupture process of the molecular chain of rubber. As viewed from energy, the frictional work of rubber W_f should be equal to the sum of the adhesion energy of rubber surface U_{ad} and the rupture energy of the molecule chain of rubber U during the whole friction process, that is

$$W_f=U_{ad}+U \tag{2.3}$$

$$U=kW_f \tag{2.4}$$

$$U_{ad}=(1-k)W_f \tag{2.5}$$

Where, k is the proportionality coefficient, which is determined by the actual process of friction.

2.3.1.1 Analysis of the Adhesion-Shear Processes of Rubber and its Counterpart

The stiff asperity is simplified as a needle-point, and the following physical model is used to analyze the energy loss of a system resulting from the rubber adhesion during the process of friction (Fig. 2.7)

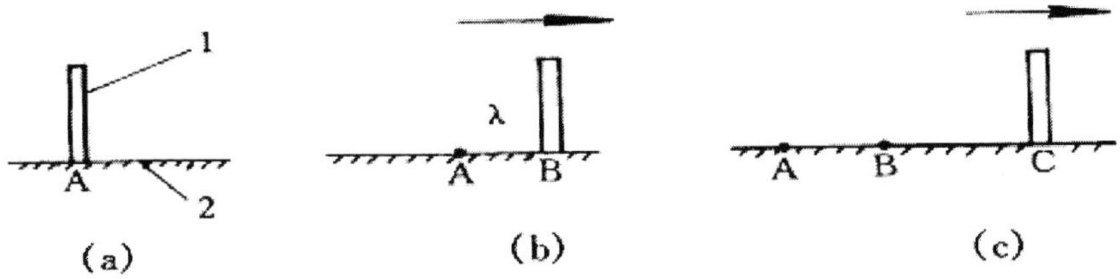

Fig.2.7. Scheme of adhesion-shear process of rubber with its counterpart
1.needle; 2.rubber

The point A on rubber surface is assumed to be adhesive with the needle-point due to the interaction of the molecules on the inter surface (Fig.2.7(a)). Soon afterwards the needle slides a distance of λ and reaches at point B on the rubber surface, namely AB=λ. In the mean time, the part of the rubber surface adhered to the needle-point at point A is also stretched to point B and produced elastic deformation, but the corresponding energy is stored at point A in the form of elastic (deformation) energy (Fig.2.7(b)). Once the elastic stress exceeds the strength of the adhesive point, this adhesive point will be sheared. At that moment, the part of the rubber stretched from point A relaxes immediately and the partial energy kept in point A returns to the system, while the rest is dissipated in heat. At the same time, the needle continues sliding forward until forming a new adhesion at point C (Fig.2.7(c)). If the maximum stress of area Δa at point A is equal to σ_{max}, the work for stretching point A will be given by

$$A_1=\sigma_{min}\lambda\frac{\Delta a}{2} \tag{2.6}$$

When the adhesion point is sheared, the energy loss of system resulted from the viscoelasticity of rubber, U_{ad}, is expressed as

$$U_{ad}=\sigma_{min}\lambda\Delta a\,\beta \tag{2.7}$$

Where β is the coefficient of energy loss.

The above coefficient can be calculated by using the Voigt viscoelastic model, namely the spring-dashpot model (Figure 2.8).

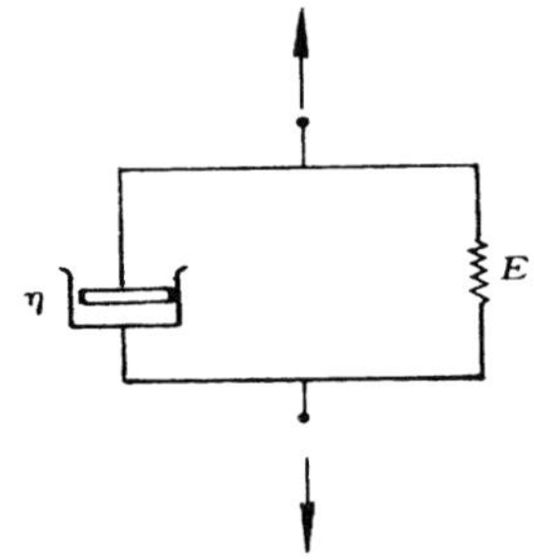

Fig. 2.8. Voigt model

When the sinusoidal vibration is added to one side of the model in quarter-period, the work for stretching spring, which equals to the elastic energy stored in the spring in the corresponding period is given as follows [19]:

$$Ae=E_mL\frac{\delta_0^2}{2} \tag{2.8}$$

Where, E_m - modulus of spring;

L - characteristic length;

δ_0 - maximum value of extension of spring.

The energy loss of dashpot in the corresponding period should be [19]:

$$E_d=\pi\eta\omega=\frac{1}{4}L\delta_0^2 \tag{2.9}$$

Where ω- frequency of vibration of dashpot;

η- viscosity of dashpot.

Whereas,

$$\beta=E_d/A_e \tag{2.10}$$

The relaxing time τ and tangent modulus $\tan\delta$ might be calculated with the following two equations respectively [19]:

$$\tau=\eta/E_m \tag{2.11}$$

$$\tan\delta=\tau\omega \tag{2.12}$$

Inserting equations (2.8), (2.9), (2.11) and (2.12) into equation (2.10), we have:

$$\beta = \pi \frac{\tan \delta}{2} \tag{2.13}$$

Taking formula (2.13) into equation (2.7), the expression of energy loss resulted from rubber adhesion is obtained:

$$U_{ad} = \pi \sigma_{\max} \lambda \Delta \alpha \frac{\tan \delta}{2} \tag{2.14}$$

$$\Delta \alpha \propto \frac{N}{H} \tag{2.15}$$

Where, H - hardness of material;

N - normal load.

Based on equation (2.5), the friction work resulted from rubber adhesion can be expressed as:

$$W_{ad} = (1-k)\, F\, \lambda \tag{2.16}$$

Where, F is the friction force.

As the energy loss U_{ad} equals to the frictional work W_{ad}, the following equation can be obtained on the basis of equations (2.14), (2.15) and (2.16):

$$(1-k)\, F = K_1\, \sigma_{\max} \frac{N}{H} \tan \delta \tag{2.17}$$

Where, K_1 is a proportional constant.

According to equation (2.16), the coefficient of friction corresponding to rubber adhesion can be expressed as:

$$f_{ad} = (1-k) \frac{F}{N} \tag{2.18}$$

Inserting equation (2.17) into equation (2.18), we have

$$f_{ad} = K_1\, \sigma_{\max} \frac{\tan \delta}{H} \tag{2.19}$$

Whereas σ_{max} is relevant to the interfacial adhesion energy, that is [19]

$$\sigma_{\max} = \phi \frac{E}{N^r} \tag{2.20}$$

Where, r - exponent, r<1;

E - elastic modulus of rubber;

Φ - function related to the interfacial adhesive behavior.

By inserting equation (2.20) into equation (2.19), the coefficient of friction for rubber adhesion can be obtained:

$$f_{ad} = K_1 \phi \frac{E}{N^r} \frac{\tan \delta}{H} \tag{2.21}$$

2.3.1.2 Analysis of the Rupture Process of Rubber Molecular Chain

The rupture process of a rubber molecular chain can be described as follows (Figure 2.9). Under the action of normal load N, the needle inserts in the rubber surface at a depth of h. While relative motion occurs, under the action of tangential load, the molecular chain of rubber is stretched out in the direction of movement. When once the stretching stress exceeds the tensile strength of the molecular chain of rubber, the molecular chain is ruptured. A further analysis of this physical process is given as below.

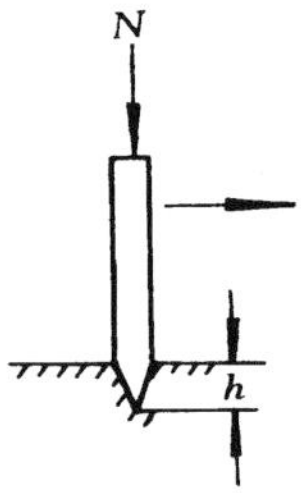

Fig.2.9. Scheme of needle inserting into rubber surface

Firstly the various factors influencing the insertion depth h of the needle into the rubber surface are examined. While the needle inserts into the rubber surface, the frictional resistance supported by the surface of needle contacting with the rubber is given by

$$F_1 = f_1 d_N S \tag{2.22}$$

Where, f_1 - frictional coefficient, during the needle against the rubber in relative motion;

d_N - normal force applied by the rubber to a micro area of the surface of needle;

S - surface area of the needle inserted into the rubber, namely

$$S = \pi h^2 \tan\frac{\theta}{2} \sec\frac{\theta}{2} \tag{2.23}$$

Where θ is the cone apex angle of the needle-point.

Because the rubber is balanced with the needle in the normal direction, the normal load N can be described as

$$N = F_1 \cos\frac{\theta}{2} \tag{2.24}$$

Substituting equations (2.22) and (2.23) into equation (2.24), we obtain

$$N = f_1 d_N \pi h^2 \tan\frac{\theta}{2} \tag{2.25}$$

Then,

$$h \propto \left[\frac{N}{\tan\frac{\theta}{2}}\right]^{0.5} \tag{2.26}$$

In accordance with the equation (2.26), the main factors influencing the insertion depth of the needle into the rubber surface h, are normal load N and cone apex angle of needle-point θ.

What follows is an analysis of the rupture process of the molecular chain of rubber under the action of tangential force.

Figure 2.10 is a sketch to show the molecular chain of rubber being stretched before rupture. ADB and ACB are assumed to represent the lengths of a molecular chain of rubber before and after stretching respectively. Thus, when the molecular chain ADB is stretched in the mid-point D (namely AD=BD), the stretching length Δx is given by

$$\Delta x = (AC - AD) + (BC - BD) = \frac{2\lambda}{\sin\alpha} - \frac{2\lambda}{\tan\alpha} \tag{2.27}$$

Then,

$$\Delta x = 2\lambda \tan\frac{\alpha}{2} \tag{2.28}$$

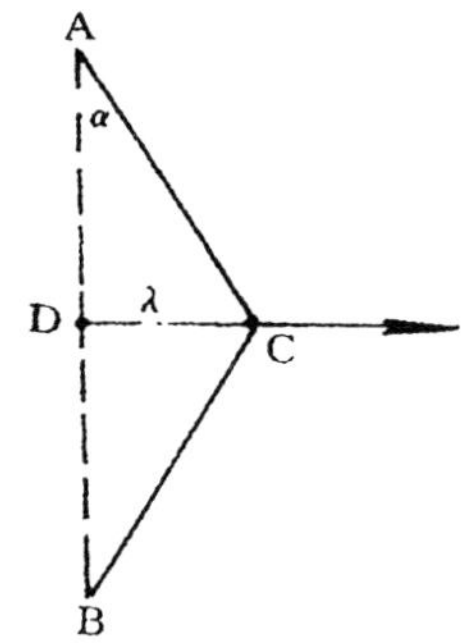

Fig. 2.10. Sketch of the stretching of a molecular chain of rubber

Assuming that the pulling force of the molecule chain, namely the tangential force being equal and opposite to the friction force F, is proportional to the stretching length $\triangle x$, then the work of a molecular chain during the whole process from stretching to rupture is expressed as

$$A_2 = \tau_{max} \frac{\Delta x}{2} \tag{2.29}$$

Where τ_{max} is the maximum tangential force to pull apart the molecular chain.

Considering that a needle contacts with n molecular chains at the same time, the sum of work exerted by a needle on n molecular chain is given by

$$A_0 = nA_2 \tag{2.30}$$

After the rupture of the molecular chain, the energy loss of system resulted from the viscoelasticity of rubber U_v can be given by

$$U_v = nA_2\beta \tag{2.31}$$

Inserting equations (2.28) and (2.29) into equation (2.31), we have

$$U_v = n\tau_{max}\lambda \tan\frac{\alpha}{2}\beta \tag{2.32}$$

In the view of energy balance, the above energy is equal to the rupture energy U, that is

$$U_v = U \tag{2.33}$$

Upon the substitution of equation (2.4) in equation (2.33), we have

$$U_v = k\omega_f \tag{2.34}$$

and

$$U_v = F\lambda \tag{2.35}$$

Assuming that the number of molecular chains contacted with a needle at the same time n is proportional to the insertion depth of the needle into the rubber surface h, i.e., $n \propto$ h. Inserting equations (2.13),(2.26), (2.34) and (2.35) into equation (2.32), gives:

$$kF = K_2\left[\frac{N}{\tan\frac{\theta}{2}}\right]^{\frac{1}{2}} \tan\delta\tau_{max} \tag{2.36}$$

$$K_2 = K_1\frac{\pi}{2}\tan\frac{\alpha}{2} \tag{2.37}$$

Where, K_1 is constant, and K_2 is also constant for a specified rubber.

From the equation (2.4), the frictional coefficient caused by the rupture of the molecular chain of rubber f' can be expressed as

$$f' = k\frac{F}{N} \tag{2.38}$$

Thus,

$$f' = K_2 \tan \delta \tau_{\max} (N \tan \frac{\theta}{2})^{-\frac{1}{2}} \tag{2.39}$$

2.3.1.3 Frictional Coefficient of Rubber by a Point Contact

In accordance with the above analysis, the frictional coefficient of rubber by a contact point f should consist of two parts, that is

$$f = f_{ad} + f' \tag{2.40}$$

Inserting equations (2.21) and (2.39) into equation (2.40), the expression of the frictional coefficient of rubber by a point contact is

$$f = \left[K_1 \frac{\phi(\frac{E}{N^r})}{H} + K_2 \tau_{\max} (N \tan \frac{\theta}{2})^{-\frac{1}{2}} \right] \tan \delta \tag{2.41}$$

From equation (2.41), it can be seen that the frictional coefficient of rubber is related to a number of factors, such as normal load, tangential load, cone apex angle of asperity as well as the elasticity, viscoelasticity, strength and hardness of rubber, etc.

2.3.2 Experimental Studies in the Friction Process

2.3.2.1 Experimental Methods

The experiments were performed using a modified Tabor-Eldredge abrader [20]. A replaceable conical steel indenter with 3 mm diameter is fixed on one side of the balanced beam of the abrader. This indenter produces friction on the surface of NBR sample in linear relative motion. The balanced beam is mounted on the bed of a small lathe driven by a small-sized motor.

In the experiments, various indenters with different cone apex angles in the range of 10°～150° were used. The normal loads applied directly to the indenter were 0.5N, 1.0N and 1.5N respectively. The sliding speeds were 7.84×10^{-6} m/s, 1.95×10^{-4} m/s, 7.84×10^{-4} m/s and 1.56×10^{-3} m/s respectively. Polydimethylsiloxane with viscosities of 9.4 mPa·s, 94 mPa·s and 950 mPa·s respectively was used as lubricant to investigate the effect of lubrication on frictional behavior.

2.3.2.2 Variation of Friction Force

The variation of friction force against time under the action of an indenter with normal load 1.0 N and cone apex angle . of 30°is shown in Figure 2.11.

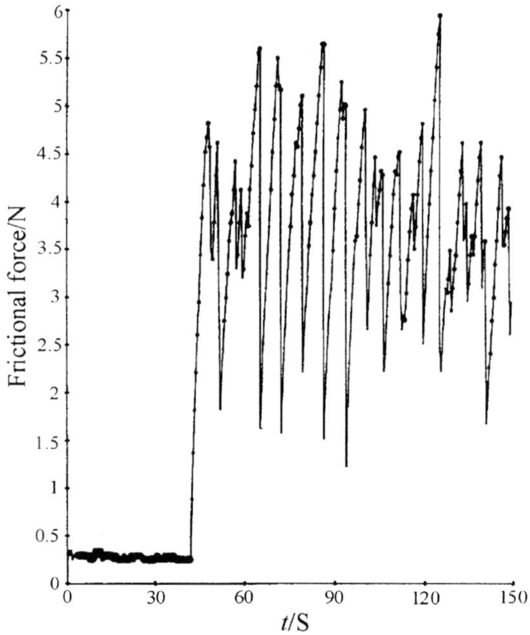

Fig. 2.11. Variation of friction force (θ=30°)

Obviously, the friction force is varied periodically with time, as shown in Fig. 2.11. According to the above theoretical analysis, it is possibly explained as that under the conditions of a point contact and this point exerted by tangential load, the molecular chain of rubber at that point is stretched along with the sliding direction, consequently, the corresponding friction force increases gradually. Once the tangential stress acting on the molecular chain at this point reaches its limiting strength, the friction force also goes its maximum, thus the molecular chain is ruptured. Next, the ruptured molecular chain of rubber recovers rapidly to the original place, in the mean time, the friction force decreases to a minimum. After that, this process takes place at another new contact point and goes on repeatedly in this way. It is why the friction force varies periodically with time.

2.3.2.3 Effect of the Cone Apex Angle of Indenter on the Frictional Characteristics

The experimental results demonstrated that the smaller the cone apex angle θ of the indenter, the larger the variable extent of the friction force is, and the fluctuation period of the friction force is longer. Possibly, the reason is that the smaller the cone apex angle, the larger the insertion depth of the indenter is, as a result, the consumption of energy in the rupture of the molecular chain is increased. At this time, the main mechanism resulted in the rubber friction should be the rupture of the molecular chain. When the cone apex angle θ is larger, so is the absolute value of the friction force, but the amplitude and period of fluctuation of the friction force are smaller. It is possibly that the insertion depth of an indenter with a big angle θ is less, and the contact area of the indenter with the rubber is larger. In this situation, the friction mechanism is mainly regarded as the adhesion between the rubber and the indenter.

The variation of frictional coefficient with the cone apex angle θ of indenter is shown in Table 2.2.

Table 2.2 Variation of frictional coefficient f with the cone apex angle θ(N=1 N)

θ	10°	30°	60°	120°	150°
f	2.8	3.5	0.73	0.24	0.22

Based on the theory advanced above, the coefficient of friction f consists of two parts. One is resulted from the action of rubber adhesion, f_{ad}, and another is ascribed to the rupture of the molecular chain of rubber, f'. From equation (2.39), it can be seen that f' is decreased with an increase in angle θ, thus the frictional coefficient f is also varied in the same way. This theoretical analysis has been verified by the experimental results listed in Table 2.2.

2.3.2.4 Effects of Normal Load and Sliding Speed on the Coefficient of Friction

The influence of normal load N and sliding speed v on the coefficient of friction under the action of an indenter with a cone apex angle of 30° is listed in Table 2.3.

As seen in Table 2.3, the coefficient of friction is increased with a decrease in normal load but with an increase in sliding speed, which proves the theoretical analysis of the influence of normal load on the coefficient of friction as expressed in equation (2.41).

Table 2.3. Coefficient of friction (θ=30°)

Sliding speed v m/s	Normal load N, N		
	0.5	1.0	1.5
7.8×10^{-6}	3.6	3.75	3.4
1.95×10^{-4}	4.3	4.1	3.7
7.8×10^{-4}	4.5	4.4	4.0

2.4. Rubber Friction by a Line Contact

2.4.1 Experimental Method

The experiments were performed on a friction and wear tester (Figure 2.12). The samples made of four kinds of rubber, namely natural rubber (NR), nitrile rubber (NBR), styrene-butadiene rubber (SBR) and polyurethane (PU) were tested under the conditions of different sliding speeds (0.06 m/s, 0.10 m/s, 0.16 m/s, and 0.20 m/s) and normal loads (8 N, 10 N and 12 N) with a line contact test-mode. The size and shape of rubber samples are shown in Figure 2.13. The razor blade was made of steel 45, and the chemical composition and mechanical properties of the steel are listed in Table 2.4 and Table 2.5 respectively [21]. Table 2.6 shows the main physical properties of four kinds of rubber. The experimental ambient temperature is 19-25℃.

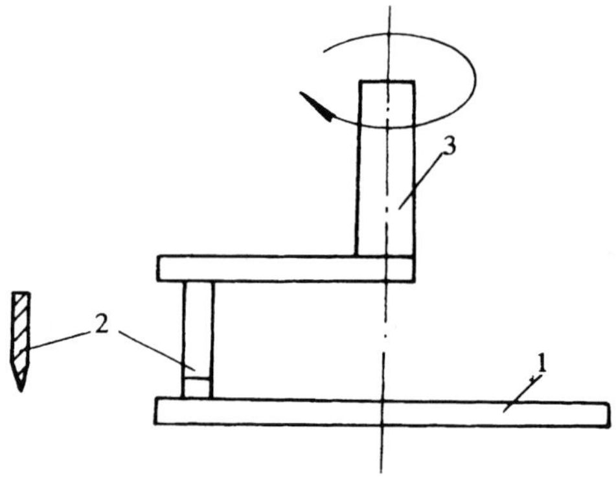

Fig. 2.12. Schematic drawing of friction and wear tester
1 - rubber sample; 2 - razor blade; 3 - driving shaft

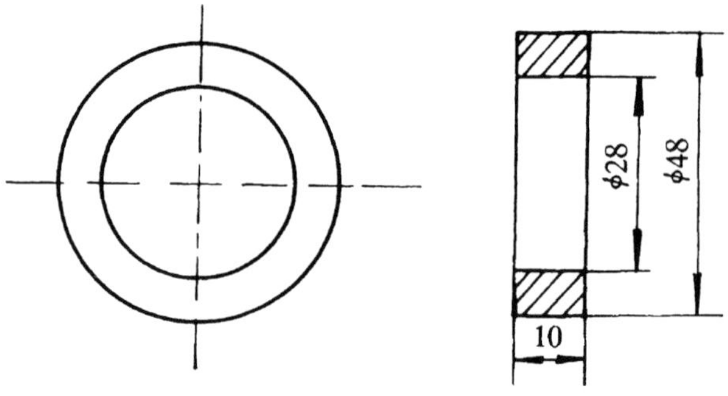

Fig. 2.13. Rubber sample

Table 2.4. Chemical composition of steel 45 (GB 699-88), %

C	Si	Mn	P	S	Cr	Ni
0.42~0.50	0.17~0.37	0.50~0.80	≤0.035	≤0.035	≤0.25	≤0.25

Table 2.5. Mechanical properties of steel 45 (GB 699-88)

Yield Strength σ_s, MPa ≥	Tensile Strength σ_b, MPa ≥	Extensibility δ_s, % ≥	Contraction Percentage of Areas ψ, % ≥	Impact Work A_k, J ≥	Hardness H_B ≤
355	600	16	40	39	197~241

Table 2.6. Main physical properties of rubber

Materials	NR	SBR	NBR	PU
Density (10^3 kg/m^3)	0.955	1.64	1.33	1.61
Tensile strength (10^4, N/m^2)	2452～3452	1471～2942	1431～1961	1370
Ratio of stretching (%)	650～900	300～800	300～800	500
Resistance of rupture	Excellent	Good	excellent	good
Wearability	Excellent	Excellent	excellent	excellent
Hardness (Shore)	20～100	10～100	40～130	13
Service temperature (℃)	100	170	80	140～160

2.4.2 Variation of Friction Force (Torque) with Time

The variation of friction torque of NBR, *M*, with time is shown in Figure 2.14.

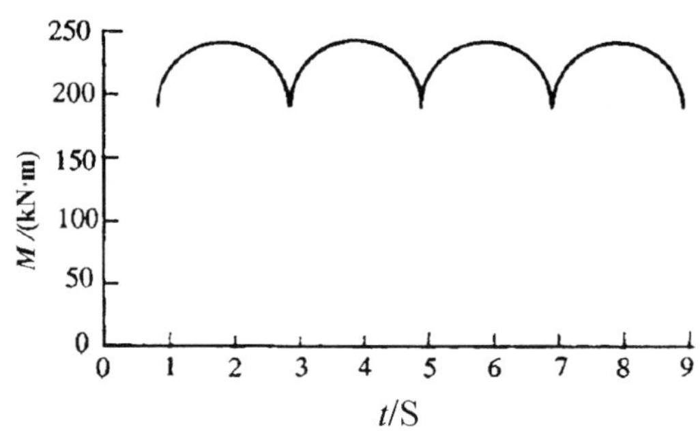

Fig.2.14. Variation of friction torque with time (N=12 N, υ=0.1 m/s)

As seen in Fig. 2.14, the friction torque varies periodically with time. Therefore, the friction force of rubber by a line contact is also changed periodically with time in the same way as that by a point contact.

The contact between the razor blade and the rubber surface might be considered as a cylindrical surface with much small curvature radius in contact with the rubber plane. The sketch of their interaction is shown in Figure 2.15. Under the action of a normal load, the contacting angle between the razor blade and the rubber is expressed as ϕ, and the corresponding friction force can be described as

$$F = f\int_0^{\varphi} rbp(\varphi)d\varphi \qquad (2.42)$$

Where, r - curvature radius of the tip of razor blade;

b - width of razor blade;

$P(\phi)$ - contacting stress function of the rubber surface.

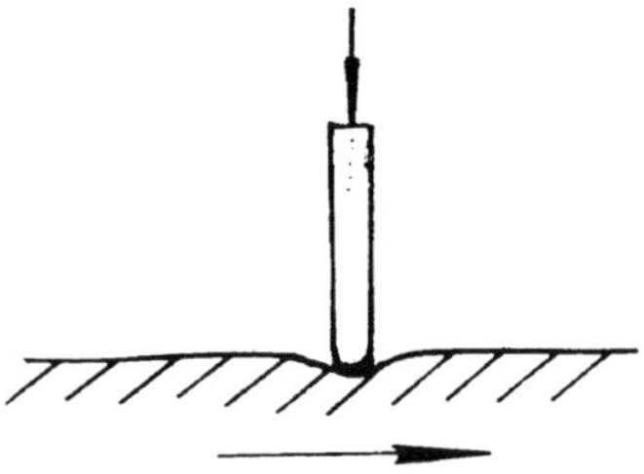

Fig. 2.15. Sketch of the rubber interacting with razor blade

Since the rubber possesses of viscoelastic property, the variety of stress is out of step with that of strain. It might be thought that the function of contact stress $P(\phi)$ is unchangeable under the condition of larger sliding speed, and it is exercised but slight influence by the variation of the contact angle. Therefore, from the equation (2.42), it can be deduced that the frictional force is increased with the contact angle.

When the razor blade contacts with the rubber and moves forward, the molecular chain of rubber is stretched in the direction of movement. As a result, the rear of the contacting zone of the razor blade with rubber is exerted by the action of pulling force. Once the pulling force exceeds the adhesive force existed in the contacting zone, the rubber begins to lose contact with the razor blade in this location. In the meantime, the contact angle ϕ decreases gradually, and then does the friction force. When the contact angle ϕ reaches the minimum, the friction force is also reduced to the minimum. Soon afterwards, under the action of normal load, the razor blade is again in contact with the rubber. Thus the corresponding contact angle ϕ is enlarged and followed by an increase in the friction force gradually. When the contact angle reaches a maximum, the friction force is also at a maximum. Therefore, the friction force (torque) varies periodically with time.

2.4.3 Influence of the Normal Load and Sliding Speed on the Coefficient of Friction

On condition that the sliding speed v is unchangeable, the variation of the frictional coefficient of four types of rubber with normal load N is listed in Table 2.7.

Table 2.7. Coefficient of friction of four types of rubber (v=0.16 m/s)

N (N)	8	10	12
NR	0.88	0.75	0.73
SBR	0.82	0.79	0.77
NBR	1.20	1.10	1.02
PU	0.55	0.53	0.48

As seen from Table 2.6, the frictional coefficient of rubber slightly decreases with the increase in normal load. These experimental results verified that equation (2.41) is also applicable for the rubber friction by a line contact.

Under the condition of a constant normal load N, the variation of the frictional coefficient of four types of rubber with sliding speed is shown in Tables 2.8 and 2.9.

Table 2.8. Coefficient of friction of NR and NBR (N=12 N)

v, m/s	0.06	0.10	0.16	0.20
NR	0.69	0.73	0.73	0.73
NBR	1.10	1.10	1.02	0.99

Table 2.9. Coefficient of friction of SBR and PU (N=8 N)

v, m/s	0.06	0.10	0.16	0.20
SBR	0.77	0.82	0.82	0.82
PU	0.49	0.55	0.55	0.55

As seen by the experimental results (Tables 2.8 and 2.9), the coefficient of friction of the rubber by a line contact is basically not changed with the sliding speed. It is possible that under the condition of a line contact, the heat-sinking condition is much better, thus the heat effect of friction does not exist generally. Consequently, the performance of rubber is stable. The value of the frictional coefficient is also maintained on the whole.

2.5 Tires Traction (Friction) Forces on Ice and Snow

2.5.1 Rubber-like Materials on Ice

Experimental study on the characteristics of friction of rubber-like materials and tread compounds on ice was probably started in the 1940's [22]. Considerable research has been made since then.

Grosch [23] ascribed rubber friction to two mechanisms: one is adhesion in the contact region for the friction on smooth surfaces; the other is plough or deformation for the friction on ice with rough surfaces. He found that under the condition of proper strain and temperatures in the range –26℃ to –40℃, ice crystallizes fast and adhesive friction decreases, but hysteresis loss is not affected. In the meantime, the friction characteristics of rubber on ice are dependent on the properties of rubber but not that of ice. Southern and Walker [24] investigated the effects of speed and temperature on the coefficient of friction of unfilled rubber vulcanizate sliding on the smooth surface of ice and compared experimental results with that obtained from the rubber sliding on the smooth surfaces of the other materials under otherwise identical conditions. In order to avoid the melting of ice, the tests were performed at low sliding speeds in the range 3×10^{-5} ms^{-1} to 0.01 ms^{-1}, and a middle contact pressure of 10^5 Pa or so. The results showed that the friction mechanism and the maximum value of the frictional coefficient are just the same for the rubber on smooth surfaces of both ice and the other materials. It might be concluded that the friction is mainly dependent on the rubber properties.

The ingenious and pioneering researches of Schallamach [25] have shown that when softer rubber slides over a hard tack, the relative motion can be due solely to the separation of the surfaces in narrow 'waves of detachment', namely so called Schallamach waves. These waves move across the contact area in the rubber. In the meantime, the contact surfaces between waves generate strongly adhering. The buckling of rubber surfaces is therefore attributed to the tangential compressive stresses in the contact area and the motive force driving the waves is a tangential stress gradient.

Based on the study of Schallamach, Roberts and Jackson [26] applied the surface energy approach to analyze the sliding friction at a rubber-track interface. As shown in Fig.2.16, a rubber slab is pulled at a speed v_s over a hard track by a tangential stress F.

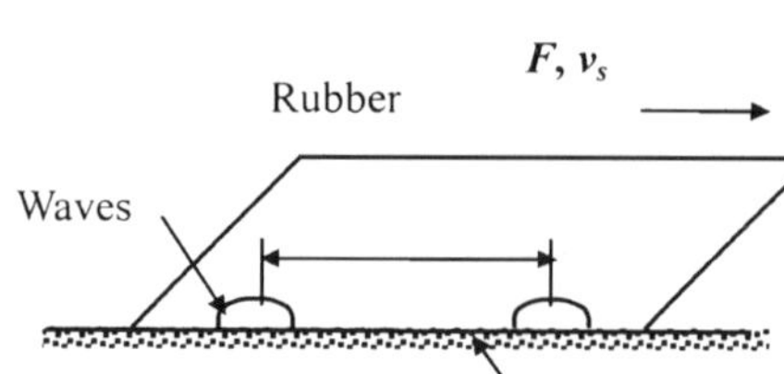

Fig. 2.16. Occurrence of Schallamach Waves in the interface between rubber and track [26]

The waves move at a constant speed and their spacing apart is λ. If the rate-dependent surface energy per unit area required to peel the rubber from the track is γ, moreover, if the internal energy loss in the bulk of the rubber is ignored, then in the steady-state sliding where the elastic energy of the rubber is constant the energy loss is solely associated with the peeling. Combining the work per unit area done by the tangential stress F in time dt, $Fv_s dt$, with the energy lost by waves, $(\gamma\ /\lambda)dt$, we have:

$$F = \gamma\omega/(\lambda v_s) \tag{2.43}$$

It was found that equation (2.43) might accurately predict correct order of magnitude for the sliding friction force of a smooth rubber contact. In some cases the integrated theoretical frictional stress is within 10% of the experimental observations, as shown in Table 2.10.

Table 2.10. Wave observations and the friction of rubber hemispheres on glass

Rubber Vulcanizates (Hardness IRHD)		Body velocity v_s/m·s^{-1}	Centre of Contact Observations /mm·s^{-1}	/mm	/J·m^{-2}	Frictional Stress (MPa) Theory Center	Theory Integrated	Observed
Natural Rubber	Peroxide cure (35)	590	17.4± 4	1.3± 0.2	5.98	0.14	0.16± 0.03	0.15± 0.02
	Sulphur cure (44)	240	5.9± 0.6	0.34± 0.1	5.01	0.36	0.53± 0.015	0.48± 0.04
PCR (48)		100	2.3± 0.5	0.62± 0.1	21.4	0.79	0.83± 0.07	0.70± 0.03
SBR (50)		110	2.9± 0.3	0.52± 0.08	9.12	0.46	0.44± 0.02	0.45± 0.03
BR (34)		50	0.59± 0.06	1.2± 0.2	15.1	0.15	0.16± 0.03	0.20± 0.03
PBR (53)		110	4.0± 0.8	0.34± 0.08	1.78	0.19	0.20± 0.03	0.26± 0.03
ABR (43)		30	0.70± 0.2	0.6± 0.1	5.76	0.22	0.22± 0.06	0.29± 0.02

In order to further understand the mechanisms of rubber friction on ice with different temperatures, Roberts and Richardson [27] investigated the interface of rubber-ice friction. A transparent smooth-surface rubber hemisphere with radius of 18.5 mm was used as the test sample to examine the transient physical phenomenon in the contact region. The ice track was supported on a glass turntable driven externally by a speed change motor. The sliding speeds can be given in the range 10^{-6} ms^{-1} to 1 ms^{-1}. Experiments were carried out inside a refrigerator, which is a 0.4 m^3 chest-type freezer capable of maintaining a temperature of –35℃. The results showed that the rubber could adhere well to cold ice. However, when thc temperature of ice nears the melting point, both adhesion and friction are nearly lost. Measurement details were shown in Fig.2.17. This experiment was conducted at the temperatures ranging from –1℃ to –32℃ and at sliding speeds in the range 10^{-6} ms^{-1} to 10^{-2} ms^{-1}. With speed greater than 10^{-3} ms^{-1} and temperature less than –15℃, the Schallamach waves usually occurred in the contact area, the friction on ice was very high. Thus good grip was indicated between the rubber and the cold ice, which was dependent on the surface interaction and the viscoelastic properties of rubber. The friction coefficient on cold ice was high at around 2 and showed no marked speed dependence. There was a sharp fall for all the concerning speeds at temperature above –10℃ because of the disappearance of the Schallamach waves and the easy flow of 'warm' ice. On warm ice, the friction coefficient at a low sliding speed was to a greater extent related to the properties of ice other than the properties of rubber. Only at intermediate and high sliding speeds as well as before the friction melt of ice being predominant could the properties of rubber be considered [28].

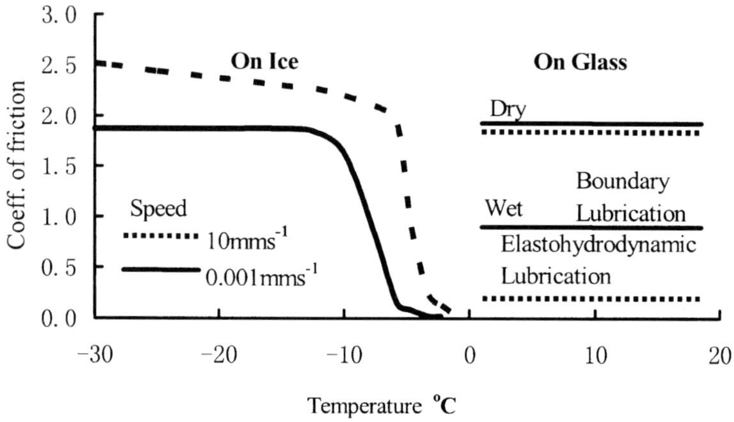

Fig 2.17. Sliding friction of rubber on ice and glass [27]

Sliding at a low speed on ice warmer than –3℃ results in the near disappearance of friction resistance, which may be a result of both surface water and the ready flow properties of the underlying ice. An infrared thermometer was used in the test to sense the ice track temperature a short distance behind the rubber pad. Both direct observations of the contact area and temperature measurements of the ice track give an indication of ice melting due to frictional heating other than pressure melting. The higher the sliding speed and the frictional heating quantity, the lower the friction coefficient is. Also, surface water from frictional heating was viewed through the transparent rubber sample [29].

2.5.2 Tire Traction on Ice

Experimental studies on the friction characteristics of tires on ice were generally performed outside in winter. As early as 1947, Moyer [30] made braking tests for a vehicle on ice and snow. Footit [31] and William [32] studied tire traction characteristics in winter and recommended the standardized test methods so as to improve their effectiveness and applicability. They pointed out that direct comparisons between the test results were difficult due to the unexpected scatter of the test data, which was caused by the changeability of the environmental conditions of ice and snow. Hunter [33] showed that tire friction resistance on ice was affected by many factors, such as the kind of instrument, the type of available braking force, the contaminative extent of the roadway and numerous other physical factors. To examine the effect of anti-lock brakes with summer tires, all-season tires and winter type tires on ice and snow, Eddie [34] carried out tests on a flooded frozen track and on asphalt pavement with newly fallen snow. The friction values obtained were compared with those reported by other investigators. He found that the most important factors for braking and traction on snow or ice were not the tire type or tread but the temperature of the tire/snow or tire interface in the test. Surface roughness was also a significant factor. Francis, Michael and Connie [35] summarized the number of probable parameters, which directly affect the friction characteristics of tires on winter pavement as follows:

- road surface
  - -gradient
  - -bare, wet, dry
  - -temperature
  - -ice, snow
  - -snow depth and density
- vehicle type
  - -weight
  - -torque, braking force
  - -anti-lock brake, traction control
  - -tire tread depth and pattern and the number of drive axles
- aggregate
  - -size, shape
  - -application rate
  - -dispersal from passing vehicles
  - -additives, and salt

Martin and Schaefer [36] investigated the variation of the tire-road frictional coefficient with the dispersion rate of applied sand. Tests were conducted on surfaces including bare asphalt, black ice, ice and snow, ice and snow with a variety of sand overlays, ice and snow with a layer of fresh snow, and glare ice at temperatures ranging from –4℃ to –42℃.

Characteristics of tires on ice and snow under the conditions of different temperatures, velocities and surface roughness are usually evaluated on an ice/snow-covered proving ground by actual out door tests or using a tire test trailer. However, because of unstable ice/snow conditions and/or varying operating conditions, little or no repeatability can be obtained. Therefore, indoor testing acquires greater importance.

Hayhoe [37] measured the frictional coefficients of locked wheel sliding and driving traction made with an ASTM E524 smooth-tread standard test tire on an ice track at –7℃ and presented a mathematical model for estimating the traction forces of a tire in the contact region.

Shimizhu and co-workers [38-40] have developed an iced drum tester in which the drum inside is coated uniformly with ice. The tester is installed in a cold room, in which the temperature can be adjusted from –20℃ to 0℃. The tire and the drum can be controlled in constant speed mode or constant torque mode; therefore, the friction characteristics under various slip ratios can be tested. The lateral traction forces and longitudinal ones were measured at different slip angles, from 0° to 4°, and different temperatures, from -15℃ to 0℃. The effects of micro-texture and macro-unevenness of the ice surface on tire traction characteristics were also investigated. The results showed that the characteristics exhibit a remarkable dependency on ice temperature on the dense and smooth ice, but this dependency is much reduced on the ice surface with macro-unevenness.

Peng et al. [41, 42] developed an ice/snow bench tester, by which six components of the tire forces and movements and the surface temperatures of both tire and track can be recorded automatically. By experimental studies, they suggested that the inflation pressure should be lowered to a certain extent to improve the stability and controllability of a vehicle on ice or snow. They also presented a mathematical model with higher accuracy for predicting tire traction on ice.

As stated above, the most important factor affecting tire traction characteristics is the temperature of ice or snow, and then tire type, tire tread pattern, tread surface roughness and road aggregates, as well as surface macro-unevenness of ice and packed snow. Since the mechanisms of rubber-ice friction is still obscured, it is necessary to develop a suitable thermo-dynamic model for predicting tire traction and braking forces considering the combined effects of properties of both ice/snow and tread rubber, and ice/snow temperature as well as surface roughness and surface texture.

Improvements of tire traction characteristics on ice and snow are very important for vehicle handling capability and safety. It requires not only understanding the friction mechanisms of tires on ice and snow for improving traction characteristics from tire construction sizes and predicting tire traction at different test conditions, but also developing new type of tread fillers and studying new technology of polymerization for tread to make studless tires and all-season tires with the most suitable rubber. Besides, developing intelligent control systems also deserves special attention [43].

Chapter 3

AN INTRODUCTION TO WEAR

3.1 Definition and Essential Characters

3.1.1 Definition

A widely received definition of wear is still absent on account of the complexity of wear phenomena. For example, in *Encyclopedia of Tribology* the term "wear" was defined as "undesirable continuous loss of material from one or both of the surfaces of mating tribological elements due to relative motion of the surfaces" [6]. While Kragelsky and Alisin [7] held that wear of materials is a process of surface destruction in rubbing solids, which results in reduced dimensions of parts in a direction perpendicular to the rubbing surface. However, the above definitions are not very strict.

Based on the long years of research and teaching activities in the field of tribology, the author of this book intend to put a more exact definition on "wear" as follows: wear is a phenomenon of material loss generated gradually on the rubbing surfaces of solid due to a combined action of several effects, including mechanical, mechanochemical, thermochemical and electrochemical, which is mainly appeared as the continuous changes of size and/or shape of the solid surfaces. Thus it can be seen that wear phenomenon is not a single physical process caused by the mechanical action, but always accompanied by some processes of chemical reaction. The above definition explains three main features of wear phenomenon:

(1) Wear is a surface phenomenon occurred on the solid surfaces. Therefore, a typical fracture or fatigue damage caused by inner cracks of the solids does not belong to the category of wear.

(2) Wear is a phenomenon produced on the rubbing surface of the solids. It means

that a relative motion must be in existence on the contact surfaces. Pure corrosion and aging of rubber surfaces all result from chemical reactions (including oxidization) generated on the static surfaces. Therefore, these phenomena also do not belong to the category of wear.

(3) Wear inevitably results in material loss (including material transfer). Moreover, it is a gradually progressive dynamic course with a characteristic of time-dependence. Therefore, pure plastic deformation without loss of material does not belong to the category of wear too.

3.1.2 Essential Characteristics

Wear is possessed of three essential characteristics as follows:

(1) Wear is almost an inevitable phenomenon in the normal operation process of machine parts. So long as the wearing value (or wear rate) does not go beyond allowed value within the allotted operating period, it can be regarded as an allowable normal wear phenomenon during operation.

Generally, a typical wear process of machine parts can be divided into three stages as follows and as shown in Figure 3.1:

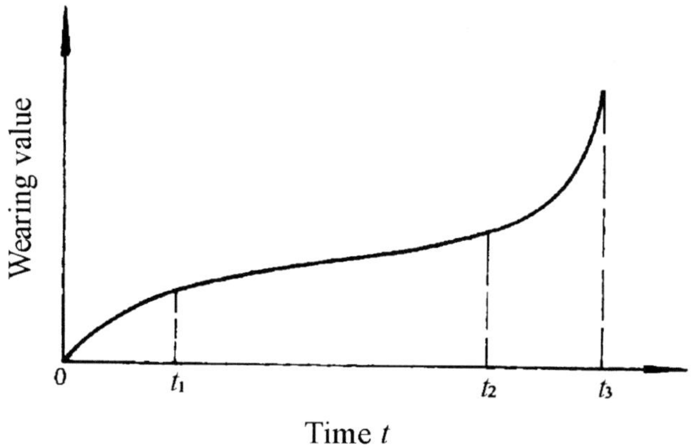

Fig. 3.1. Typical wear curve of machine parts

① Running-in (wear-in) stage (sector 0~t_1 in Figure 3.1)

On the fresh rubbing surface machined of a new friction pair, since the height of the microcosmic asperities with shaper peak is not uniform, the real contact area of the surface is small. Therefore, before putting into operation formally, it is necessary to increase loading gradually for wear-in, so as to enlarge its real contact area and avoid damaging the mating surfaces. In this stage, the wear rate increases quite quickly at the beginning and slows down gradually later on until the stable wear stage operated normally is reached. Thus this phase may be also termed unsteady-state wear stage.

② Normal (steady-state) wear stage (sector t_1~t_2 in Figure 3.1)

In this period, the wear rate is approximately kept constant and the frictional surfaces are worn slowly.

③ Accidental wear stage (sector t_2~t_3 in Figure 3.1)

After the component operated for a long time, its tolerance clearance increased. Thus its precision and performance go down, and the lubrication condition is worsened. As a result, the wear rate increases suddenly. The performance and efficiency of the machine decrease obviously. Moreover, abnormal noise and vibration often occur and the temperature of the friction pair goes up rapidly. Finally, the machine part completely fails.

(2) Wear is a typical nonlinear dynamic process. It also consists of three stages corresponding to the above three wear processes, namely,

① Self-organization stage

In this stage, the morphology of a couple initial surfaces of a friction pair forms spontaneously matchable and suitable state each other through the agency of continuous adjusting and amending. This is a process of unceasing learning, adapting and revising.

② Chaos stage

In this phase, an ordered stable texture comes into being on the worn surfaces of the friction couple. Moreover, the frictional coefficient and friction temperature are roughly constant. It means that the frictional pair reaches a smooth, steady and ordered state.

③ Instability stage

In this period, the frictional pair is unstable and loses its functions completely.

(3) Wear is not only an expression of a material's own intrinsic property, but the reflection of the feature of a tribo-system as well.

This kind of characteristics can be expressed by the following function [2]:

$$w = f(x, s) \tag{3.1}$$

Where, x - operating parameters, including load, speed, temperature, operating duration, moving distance and type of the relative motion etc.;

s - structure of tribo-system, it is represented by the following set:

$$s = \{A, P, R\} \tag{3.2}$$

Where, A - a set of elements (including environment) of the tribo-system;

P - a set of the relevant properties of the elements;

R - a set of correlations between the elements and/or interrelations between the elements and the tribo-system;

In practice, one and the same component can display different wear rates or even different kinds of wear in distinct tribo-system. Even in the very same tribo-system, different operating conditions also lead to the above results. Therefore, when dealing with the wearing problem of a component, it is necessary to consider all-sidedly the characteristics of the tribo-system where the component exists in, only that can we make a precise judgment and correct analysis for its wear phenomenon. Sometimes, we can improve the wear state of one or both sides of the friction pair without changing the operating conditions, but varying the structure of the tribo-system, such as providing better lubrication conditions or environment. Thus it can be seen that wear is also possessed conditionality and relativization.

The essential parameter to characteristize the wearability is wear rate, which can usually be expressed as the following three criteria [13]:

① linear wear rate

$$R_1 = \frac{h}{L} \tag{3.3}$$

②Volumetric wear rate

$$R_v = \frac{\Delta V}{LA_a} \tag{3.4}$$

③Gravimetric wear rate

$$R_w = \frac{\Delta W}{LA_a} = \rho R_v \tag{3.5}$$

In the above three formulas, h, V, W represent the thickness, volume and weight of the removed material respectively; L is the sliding distance; A_a is the apparent contact area; . is the density of removed material.

In a few literatures, some other parameters expressed wearability can be seen as follows [13]:

① Abrasion factor

$$\lambda = \frac{\Delta V}{NL} = \frac{R_L}{p} \tag{3.6}$$

Where, p — normal pressure;

N — normal load.

② Abradibility (Energetic wear rate)

$$\gamma = \frac{\Delta V}{FL} = \frac{\Delta V}{fNL} = \frac{\lambda}{f} \tag{3.7}$$

Where, F — frictional force;

f — frictional coefficient.

③ Coefficient of abrasion resistance (wearability)

$$\beta = \frac{1}{\gamma} = \frac{f}{\lambda} \tag{3.8}$$

④ Coefficient of wear (wear constant) [44]

$$K = \frac{WH}{Nvt} \tag{3.9}$$

Where, W — wearing value;

H — hardness;

v —speed;

t — time.

The coefficient of wear represents the relation between wearing value and operating condition for a specified material. If the values of load and speed are known, and the coefficient of wear on this operating condition could be obtained, the wearing value will be estimated so as to predict the working life of the components. As various types of wear have different coefficients of wear, the wear type of a component can be also determined according to its value of wear coefficient.

⑤ Wear speed (wearing intensity)

$$I = \frac{\Delta V}{t} \tag{3.10}$$

or

$$I = \frac{\Delta W}{t} \tag{3.11}$$

In addition, the concept of relative wear-resistance is used too, which is the ratio (%) of wear rate of a standard sample to that of the sample being tested.

3.2 Classification

For more than half a century, many scholars have put forward various classification methods from various angles. However, a generally recognized methodology has not been achieved yet up to now. Not even the generally acknowledged terms and definitions for some wear phenomena have been reached yet. This situation indicates that wear, as a field or discipline branch, is much immature. It is short of a deep understanding of the varieties of complex wear phenomena and the mechanisms, and even to the extent that some wear phenomena (type) was confounded with their wear mechanisms (essence). Consequently, the specious situation and a cloudy conception of the classification of wear occurred. In view of this fact, a more commonly used classification method is adopt in this book, which was put forward by Burwell according to the wear mechanisms [45], namely, the types of wear are classified as follows:

(1)adhesive wear (adhesion);

(2)abrasive wear (abrasion);

(3)fatigue wear or surface fatigue wear;

(4)corrosive wear or tribochemical wear;

(5)the other, such as erosive wear (erosion) and a kind of wear particular to such high elasticity materials as rubber and rubber-like materials, namely frictional wear or wear caused by roll formation.

The abrasive wear, fatigue wear and frictional wear of rubber are all generated on the rigid matrix surfaces in relative motion, which are dependent on the roughness of the rigid matrix surface to a great extent. However, the abrasive wear and fatigue wear of rubber usually emerge on the rough surfaces, but the frictional wear frequently occurs on the smooth surfaces with high frictional coefficient. So far as the severity of wear, the abrasive wear and frictional wear are most serious, but fatigue wear is less. Moreover, it should be pointed out that the wear mechanism being closely correlated with the friction mechanism for the three types of wear as shown in Figure 3.2 [46, 47], is an important characteristic of rubber wear.

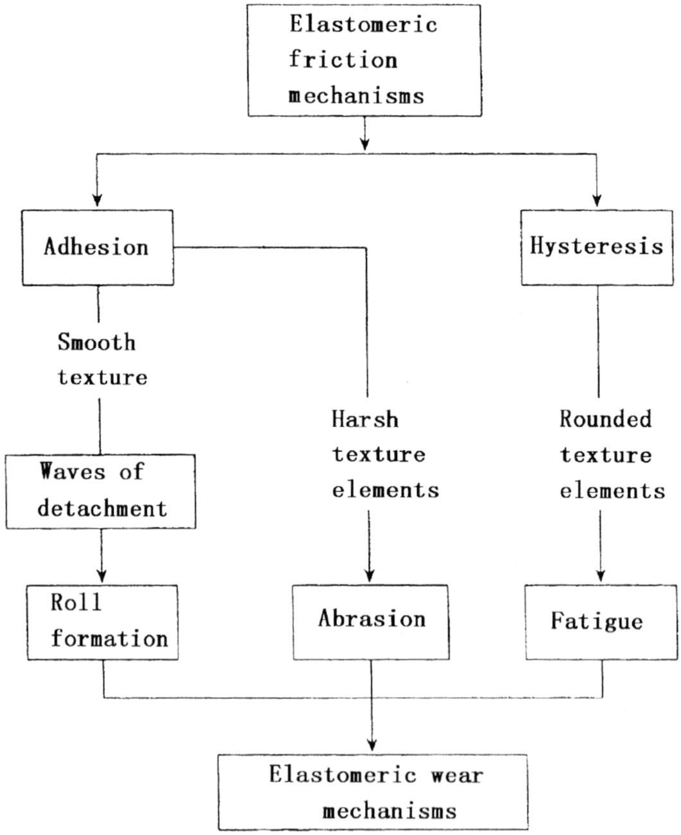

Fig.3.2. Schematic diagram of the friction and wear mechanisms in rubber-like materials [47]

Chapter 4

ABRASION

According to the difference of the morphological characteristics of worn surface, the rubber abrasion is classified into two categories: pattern and intrinsic abrasion. The basic feature of the formal is sets of parallel ridges appeared on the surface of the sample and at right angles to the sliding direction. These ridges are called abrasion patterns. The abrasion patterns of natural rubber (NR) are illustrated in Figure 4.1. Pattern abrasion is generated under the condition of unidirectional relative sliding. When the direction of relative motion is changed periodically, intrinsic abrasion occurs without abrasion patterns [17, 18]. Under otherwise identical conditions, the wear rate of intrinsic abrasion is lower than that of pattern abrasion. Figure 4.2 gives a comparison of the wear rates between the two kinds of abrasion when an unfilled NR mix slides on the silicon carbide paper [48, 49]. Rubber abrasion is also classified in two groups, namely dry abrasion and wet abrasion or hydro-abrasion in accordance with that whether a certain liquid exists on the frictional surface. However, under the condition of dry abrasion and small frictional work, a sticky layer will occur on the frictional surface owing to the surfacial interaction. This kind of abrasion might be called oily abrasion [50]. The three types of abrasion above will be described as follows.

4.1 Dry Abrasion

4.1.1 Point Contact Abrasion

In the early fifties, Schallamach [18,48,51] conducted experiments to simulate the rubber surface abraded by a single abrasive grain. He used a sharp needle and a small hemisphere of 1 mm in diameter to imitate the extreme sharpness and roundness of asperities on rough surface of solid respectively, namely sharp and blunt peaks, which can represent various shaped asperities on rough surface in broad term.

Fig.4.1. Abrasion pattern of filled NR ($\overline{F} = 470N/m$, i=500r)

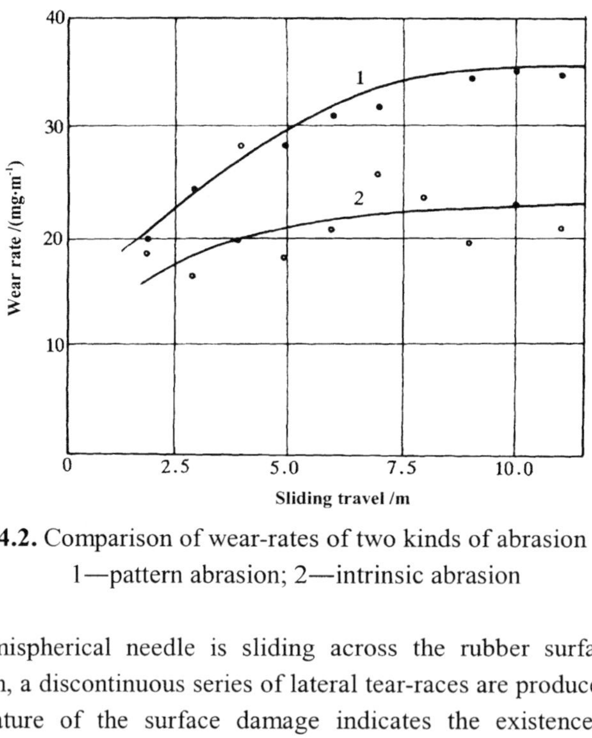

Fig.4.2. Comparison of wear-rates of two kinds of abrasion [49]
1—pattern abrasion; 2—intrinsic abrasion

When a hemispherical needle is sliding across the rubber surface under larger loading condition, a discontinuous series of lateral tear-races are produced on the surface. The periodic nature of the surface damage indicates the existence of a stick-slip mechanism during sliding. The rubber adheres locally to the hemispherical point and is stretched in the direction of travel until the elastic restoring force exceeds the sliding friction force, and then the rubber snaps back. The spacing of the tear traces in the direction of travel and the size of each tear increase as the hardness of the rubber decreases. In order to understand further the stress distribution in the rubber, which is responsible for producing this characteristic tearing process, it is studied in a cylindrical slider traveling across the transparent rubber with low sliding speed by using photoelastic technique. The photoelastic stress distribution in transparent rubber adjacent to the area contacted with the slider is shown in Fig.4.3 [13,17,19]. A stress concentration occurred at the rear of the contact area is indicated by the close spacing of the isochromatic fringes, and the rubber is in tension at this point. Any failure due to repeated sliding would therefore be expected to consist of the opening of tears in a direction at right angles to the direction of travel, as a result of this stress concentration effect [13].

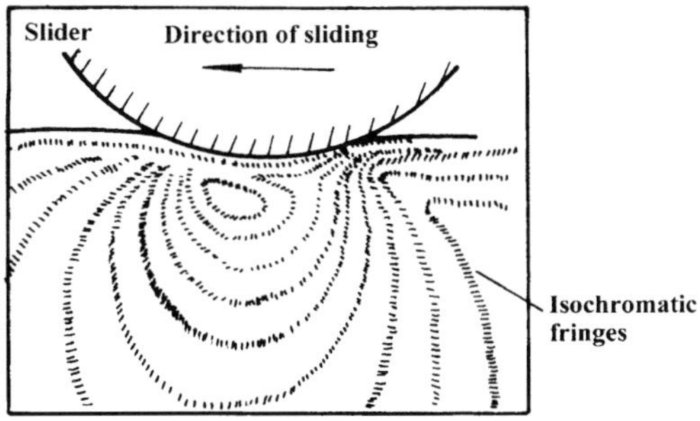

Figure 4.3. Photoelastic stress distribution in transparent rubber [13]

When the rubber surface is scratched by a sharp needle, the tractive force is larger than those in the case above, as a result, the damage of the rubber surface is more severe. For the reason that only one passage of the needle-point over the rubber track may be necessary to produce tearing and subsequent detachment of rubber particles. Figure 4.4 shows two continuous deformation stages of a rubber surface under the action of needle. As shown, the distortion of reference lines originally equidistant from each other and perpendicular to the direction of relative motion is produced, which in turn indicates the stress concentration in the vicinity of the needle point. Although the stress concentration in front of the needle is highest, frictional adhesion between rubber and needle in this location prevents rubber from tearing. Therefore, the rubber tears instead at the pinot where it first loses contact with the needle, and the tears develop laterally as indicated by the dotted lines in Fig.4.4 [13], namely the direction of tearing is perpendicular to that of the maximum stress. The pitted damage occurs on the rubber surface adjacent to the location of tearing after the needle scratches across the rubber surface. While the hardness of rubber is smaller, the depth of the pits formed on this rubber track is bigger, and the spacing between the pits is larger as well as the distribution of pits is also much uniform. With the hardness of rubber increases, the distance between the pits also decreases gradually due to the elastic deformation reducing, with the result that a continuous slot is produced finally. Schallamach [18] has pointed out that the chemical composition of the rubber itself (whether filled or unfilled) appears to have no effect on the mechanism of rubber abrasion for sharp aspirates and the primary cause of abrasion is the local stress concentration produced by fractional adhesion and mechanical interlocking between the asperity peaks and the rubber track for very sharp asperities.

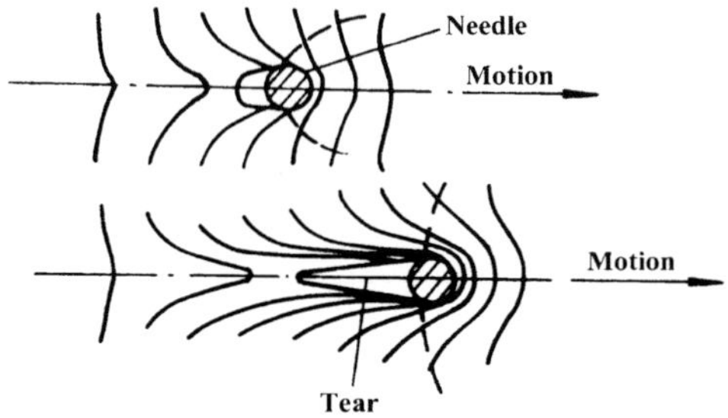

Figure 4.4. Sketch of two stages in the deformation and tearing of a rubber surface by a needle [13,19]

Moore [13,19] presented a simplified theory of rubber abrasion by a point contact and assumed that:

(1) The length of a tear is proportional to the width of contact between the abradant (whether an abrasive particle, needle or hemisphere) and the rubber.

(2) The volume of the detached rubber particle is proportional to the third power of length of the original tear.

While an abrasive particle indents into and moves along relatively to a rubber track, and let R denotes its mean radius of curvature at the tip and a denotes the mean radius of the area of contact (see Fig.4.4), then

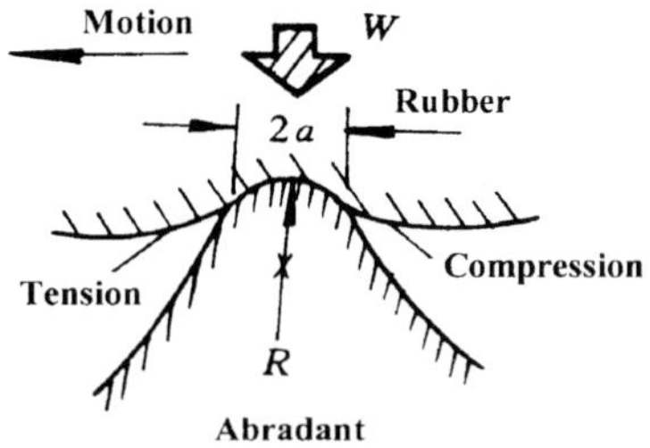

Fig.4.5. Interaction of rubber and an abradant during sliding [13]

$$a = \Phi(R, W, G) \tag{4.1}$$

Where, Φ—functional dependence;

W—applied load on the abrasive particle, which causes the particle to form a width of contact being $2a$ on the rubber surface;

G—elastic modulus of rubber. Its effective value is not the same as that of the bulk material, since the top layer near an abraded rubber surface is generally much softer then the bulk rubber due to the repeated large deformations and microcutting occurring there.

Two dimensionless parameters $\frac{a}{R}$ and $\frac{W}{GR^2}$ may be formed and then can be related by the following equation:

$$\frac{a}{R} = C_1(\frac{W}{GR^2})^{\alpha} \tag{4.2}$$

Where, the constant C_1 and the exponent a must be determined experimentally, in general, $a \approx 1/3$.

Let there be n^2 abrasive particles per unit area, and let the normal load per unit area be given by

$$N = \sum W = n^2 W \tag{4.3}$$

Then according to equation (3.6) and the assumptions above, the abrasion factor can be obtained:

$$\lambda \propto n^2 a^3 \quad 4.4)$$

By substituting for n^2 from equation (4.3) and for a from equation (4.2), we obtained:

$$\lambda = c_2 \frac{NR}{G} \quad 4.5)$$

Where $c_2 = c_1^3$=constant, the relationship above was confirmed by the experiments of four different compounds of natural rubber [18].

According to equation (3.8), the coefficient of abrasion resistance is given by:

$$\beta = \frac{f}{\lambda} = c_3 \frac{Gf}{NR} \tag{4.6}$$

Where $c_3 = \frac{1}{c_2}$=constant.

As seen from the equation (4.6), a sharp abrasive particle or a large normal load can increase the wear value of rubber. However, for the rubber with a higher elastic modulus or a larger frictional coefficient, its wearability is much better.

In the early 1980's, Ahman and Oberg [52] investigated the physical process of rubber abrasion by a point contact in the sample room of the scanning electron microscope, by using a hard alloy needle point in the shape of paramid with a cone apex angle of 24°. The samples arė made of filled NBR. Experiments were conducted under the conditions of constant velocity of 3.0×10^{-5} m/s and changing normal load of 0.2N, 0.4N and 0.17N respectively. During experiment, a crack with nucleus produced in front of the needle-point under the action of stresses. In the mean time, there was a sudden drop in the tangential stress. The wear mechanism is still cutting although no cuttings or peeling of material emerges. If these cutting processes repeat many times, the rubber would adhere to the front of needle point and wind into a roll to be separated off. At this very moment, its wear mechanism is not cutting any longer but tearing.

Late in 1980's, using a modified Tabor-Edredge friction testing machine, an experimental study in the physical process of nitrile rubber abrasion by a point contact was carried out by the present author at the Imperial College in England. The experimental method in detail is given in Chapter 2, Section 2.3.2. During experiments, each sample was cut once and away. It has been proved that the mechanism is still tearing. In general, there is no material detached from the rubber surface. However, a bump of rubber material was found at the end of each pit (Figure 4.6).

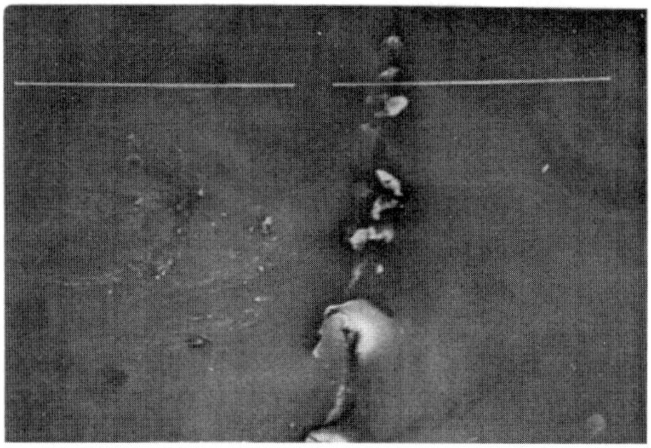

Figure 4.6. Tear on the surface of NBR (dry friction, cone apex angle 10°)

P.D Evans also reached the similar results based on his experiments conducted for natural rubber (NR) and styrene-butadiene rubber (SBR) using the same testing machine (as shown in Figures 4.7 and 4.8).

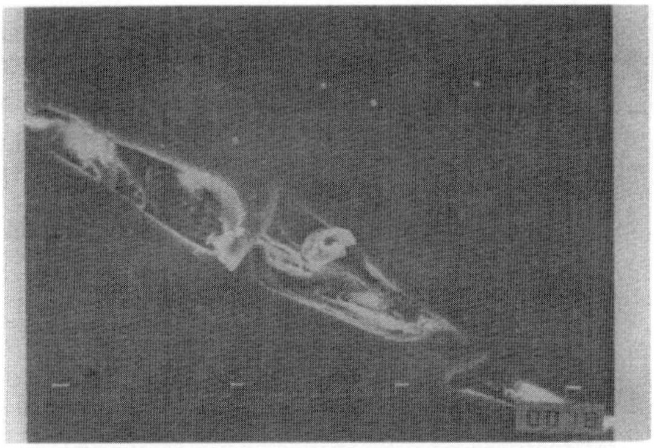

Figure 4.7. Tear on the surface of NR (dry friction, cone apex angle 60°)

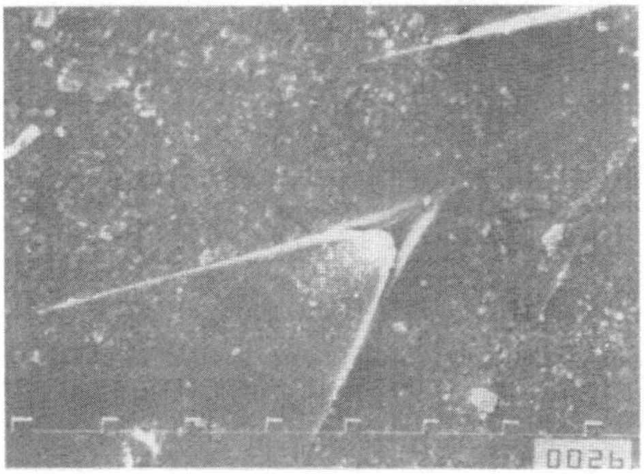

Figure 4.8. Tear on the surface of SBR(dry friction, cone apex angle 45°)

Comparing Figure 4.6 with Figure 4.7 and Figure 4.8, it can be obviously found that the feature of the pits is closely related to the hardness of rubber. The hardness of NBR material is larger, the pits on the rubber surface are nearly connected each other. As for the NR and SBR materials, there are existed a certain distance between the pits on the surface. Moreover, comparing Fig. 4.7 with Fig. 4.8 it can be further observed that for this two kinds of rubber under the condition of almost identical hardness, the larger the normal load or the smaller the cone apex angle is, the larger the spacing of the pits well be, which is consistent with the results reached by Schallamach [51].

4.1.2 Line Contact Abrasion

In consideration of that the wearing value of abrasion by a point contact is too small to measure and in order to simulate the practical pattern abrasion, Southern and Thomas [53,54] investigated rubber abrasion by a line contact in the early 1970's. They studied the mechanism of rubber abrasion with a unidirectional rotating ring-shaped sample (rubber wheel) against a stationary razor blade pressed into the sample surface in a radial direction using a blade abrader designed by themselves. They considered that the rate of movement of the pattern across the surface is closely related to the crack growth behavior [54].

Later on, Gent and Pulford [55] did a further research work on this aspect at the Institute of Polymer Science in the University of Akron, using a modified arrangement based on the blade abrader above, as shown in Figure 4.9. On the basis of experimental studies, Gent and Pulford have clearly clarified the abrasion mechanism of elastomers by a fracture process. They consider that the rubber surface develops a fine texture on two scales. One is associated with the removal of particles a few microns in size. It occurs randomly over the surface and appears to be an inevitable feature of the fracture of rubber by friction. The other, on a coarser scale, is confined to the ridges of the abrasion pattern which determines the size of the debris and probably the rate of wear also. Moreover, these general features of rubber wear by a blade abrader are quite consistent with those observed in more complicated wearing situations, such as occur in actual tire use.

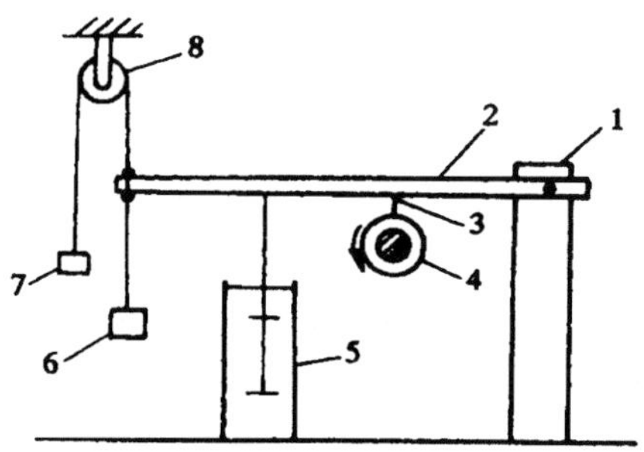

Fig.4.9. Schematic drawing of the line-contact abrasion tester [55]
1-bracket bearing; 2-beam; 3-razor blade; 4-rubber wheel;
5-dashpot; 6-load; 7-balance; 8-pulley

In the early 1980's, the present author conducted a systematical study on the mechanism of the whole process of rubber abrasion by a line contact, including two stages of unsteady and steady states, in the Institute above [56-58].

4.1.2.1 Wear Mechanisms

According to the research results [56,57], regardless of the wear stage being steady or unsteady state, the primary feature of rubber abrasion by a line contact is generated ridgy abrasion pattern on the worn surface and micro wear debris (Fig. 4.10), which indicates that the main physical process of wear includes the following two aspects.

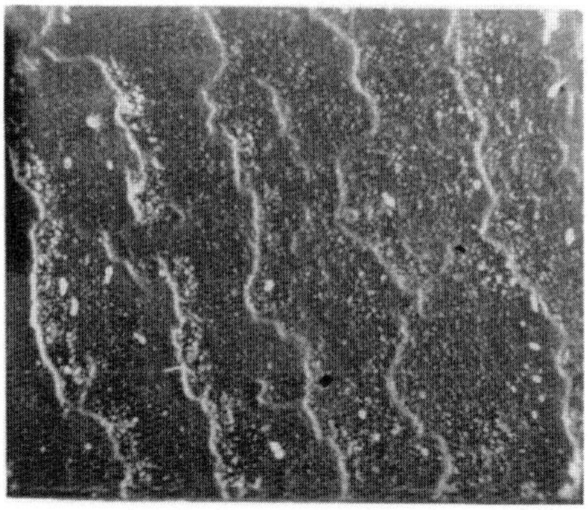

Fig.4.10. Worn surface of filled NBR ($w = 0.3kJ/m^2$)

(1) Macrodelamination

The texture of rubber is generally a micro layered or honeycombed. Under the repeated action of normal and tangential forces applied by the metal blade, two processes occurred. One is the tears of surface layer including crack growth and formation of tongue periodically produced on the rubber surface, and the other is the rupture of tongue tip resulted from the tensile stress (Figure 4.11). Consequently, the rubber surface abraded gradually in the form of delamination and shaped into ridgy abrasion pattern with sharp top, which leads the section of worn surface to a sawtooth profile. In the two processes above, the latter is the direct cause accounting for the loss of material.

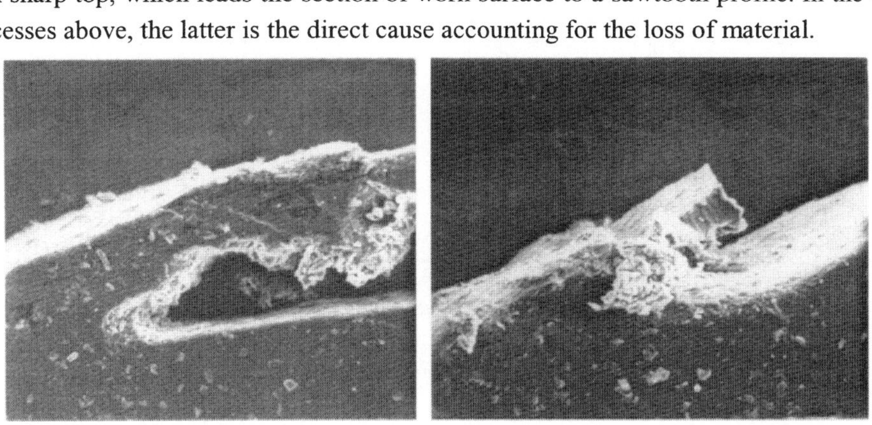

Fig.4.11. Formation of ridge on the worn surface of filled NBR (cross section), w=1.5$kJ/m2$;

(a) tongue formed; (b) tongue ruptured;

Based on the study in abrasion of natural rubber [56], it has been found that the spacing of abrasion pattern is increased with increasing frictional force, which is in keeping with the result of Reference [54]. Moreover, the pattern spacing is also under the influence of the test temperature and the glass transition temperature of the rubber [54]. However, this spacing is increased with the frictional distance too before reaching the steady-state wear stage. Therefore, the relationship between the spacing S of abrasion pattern and the frictional force F is not a linear but an exponential function as follows (Figure 4.12):

$$S = 0.03\exp(4.8\bar{F}) \qquad \bar{F} = F/h \tag{4.7}$$

Where, F—frictional force;

h—width of tongue.

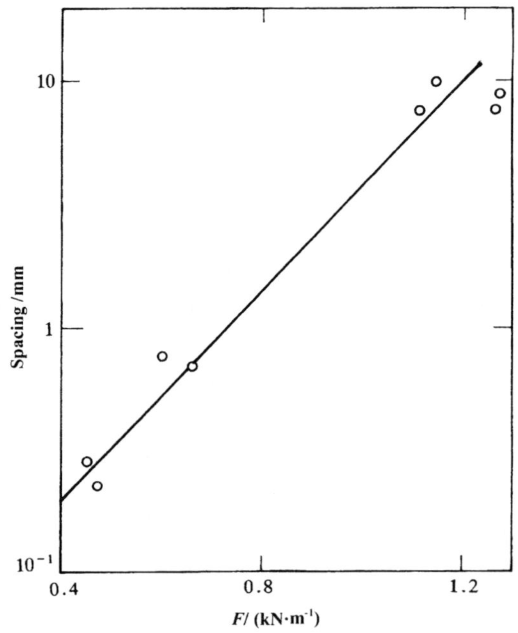

Figure 4.12. Spacings of abrasion pattern plotted against frictional force

Under the experimental condition of frictional work within the range of 0.8~1.4kJ/m2, according to the further observation and analysis of the generating process of the NBR abrasion pattern, it has been found [57] that a number of primary patterns with small spacing were formed firstly (for unfilled NBR, S≤0.5~0.7 mm, for filled NBR, S≤2~3 mm). Along with the increasing in frictional distance, the spacing also increased step by step. After reaching the wear stage of steady state, the fully developed and relatively stable secondary patterns (S=4.0~8.0 mm) were formed.

Afterwards, the spacing almost did not change with the frictional force and the processes above repeated themselves in cycles. Moreover, the primary and secondary patterns simultaneously existed and have self-similar features, namely similar characters of micro-morphology.

(2) Micromolecular fracture

Under the microcutting action of the blade, on the site of rubber surface where the macro failure is unable to happen, the micro-rapture at molecular level was occurred. It refers to some micro-particles in the range of 0.5～5 μm in dimension being detached from the worn surface. This abrasion would be arisen from the rupture of single molecules or its aggregates of the rubber as described in Reference [59], since the dimension of the micro-particles is approaching to the average stretched molecular length of NR, 2 μm or so.

Uchiyama and Ishino [60] have studied the formation mechanism of pattern abrasion of the rubber by using an unfilled isoprene rubber (IR) wheel rubbed against a steel cylinder of 6 mm in diameter and 19.7 mm in length. The diameter of the rubber specimen is ranged from 55 to 60 mm with thickness of 12 mm. The experiments were carried out at a constant sliding speed of 1 cm S-1 and different applied loads from 1.96 N to 19.6 N. Their experimental results are also proved that the spacing of abrasion pattern is proportional to the frictional force. Moreover, these abrasion patterns are continuously moved along the sliding direction during slide process. The distance of movement is linear with the travel of sliding friction. The rate of movement of the abrasion pattern, namely the magnitude of moving of abrasion pattern on the surface of rubber wheel along the sliding direction while the rubber wheel made one rotation, μm/r, is directly proportional to load and also to linear rate of wear. Obviously, the movement of abrasion pattern is resulted from the crack propagating and rupture of the root of tongue.

Of recent years, Fukohori and Yamazaki [61~63] studied the processes of rubber abrasion by a line contact, by using a steel slider of the razor blade type against the rubber block specimen which is fixed on a steel plate that moves forward and backward. The materials of the rubber specimen include unfilled and filled natural rubber, filled styrene-butadiene rubber and butadiene rubber (BR) as well as unfilled silicone rubber (SR). During experiment, the rubber specimen being in horizontality was perpendicularly contacted with the slider once and away within a reciprocating travel. The vibrations generated in the frictional sliding process were monitored by an acceleration transducer. Based on the experimental results, they considered that the essential cause to generate abrasion pattern periodically during the process of rubber abrasion is the following two periodical motions (oscillations) resulted from frictional sliding:

(1) Stick-slip motion (self-excited frictional vibration). It consists of stick and slip

stages (or states) which are generated alternatively, as illustrated in Figure 4.13(I) and (II). The frequencies are in the range of 10~20 Hz for the rubbers examined.

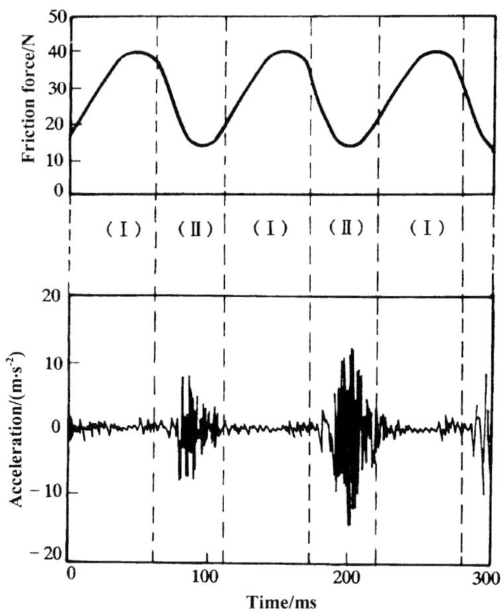

Fig. 4.13. Spectrum of frictional force and vibration (acceleration) against sliding time in filled NR [62]

(2) Microvibration. It is induced in the slip stage of the stick-slip process as shown in Fig. 4.13(II). The frequencies which agree with the intrinsic natural frequency of rubber are in the range of 500-1000Hz for the rubbers examined.

The research results (Fig.4.14) reached by Fukohori and Yamazaki [61] are proved again that the abrasion pattern is developed gradually from micro-pattern, an initially grown primary pattern with smaller spacing, to macro-pattern, a stable secondary pattern with larger spacing as stated previously in Reference [57]. Microvibration makes the micro-pattern sprouted and the stick motion makes the pattern spacing propagated. The spacing of micro-pattern in the initial stage is equal to the ratio of average sliding speed v to the intrinsic natural frequency of rubber fo, i.e., v/fo, and the final stable spacing of abrasion pattern is equal to the ratio of average velocity to the frequency of stick -slip motion Fo, i.e., v/ Fo. These results have been confirmed by experiments. Therefore, they held that two different applied forces resulted from the two kinds of motion as mentioned above produced two types of particles in different scales, namely small particles, probably of the order of 10μm or less originated from microvibration and large ones of the order of a few hundred micrometers generated by stick-slip motion.

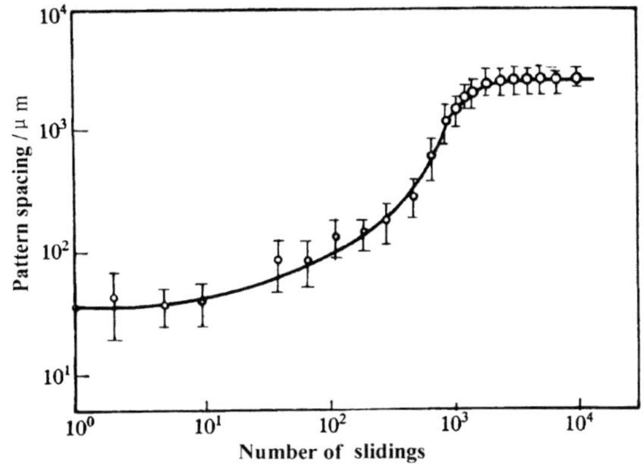

Fig. 4.14. Average pattern spacing as a function of the number of slidings (unfilled natural rubber vulcanizate) [61]

Fukohori and Yamazaki [63] pointed out further that the magnitude of the mean strain produced at the rubber surface by the two motions mentioned above is the direct reason accounted for the fracture of surface layer, namely the formation of abrasion pattern, which is directly proportional to the frictional coefficient and normal load, but is inversely as the elastic modulus of rubber. Therefore, rubber abrasion depends on the fracture resistance properties of the rubber itself to a large extent under the repeated deformation of the mean strain amplitude.

Their experimental results [62] are also demonstrated that the reinforcement by carbon black increases the frequency of both periodic motions in rubber and of course reduces the initial and final pattern spacings in rubber abrasion. In addition, the microvibration attenuates more rapidly in more filled rubber when it spreads over the rubber surface as a surface wave. Both phenomena have a great influence on the smaller abrasion of more filled rubber.

However, the wear mechanism as mentioned above is only viewed from the angle of surfacial physical effect in wear process and does not make a distinction between unsteady-state and steady-state abrasion processes. Moreover, it is also neglectful of the debris in dimension less than 10 μm. Hence, it is unable to clarify the essential cause of that the spacing of abrasion pattern increases with the sliding distance and to reveal the mechanism of rubber abrasion completely.

4.1.2.2 Linear Wear-Rate Equation

Experimental studies in the abrasion of various rubbers were conducted extensively by a number of scientists, using a line-contact abrasion tester as illustrated in Fig. 4.9 or the similar one. They all have reached the linear wear-rate equations having the same form as follows, namely the linear wear-rate follows an exponential rule [55~57, 60,64,65]:

$$R_L = k w^n \tag{4.8}$$

Where, R_L —linear wear-rate, m/r;

w —frictional work, kJ/m2;

k and n are coefficient and exponent respectively, the values of which are dependent on the kinds of rubbers, as shown in Tables 4.1 and 4.2.

Table 4.1. Coefficients ***k*** and exponents ***n*** in the steady abrasion stage at room temperature

Materials	***k***	***n***
unfilled NBR [57]	5.5×10-18	3.4
filled NBR [57]	6.5×10-15	2.1
unfilled NR [55]	5.9×10-15	2.7
unfilled NR [64,65]	1.1×10-13	2.6
unfilled SBR [55]	7.0×10-16	2.9
unfilled SBR [64,65]	2.0×10-14	2.4
filled SBR [55]	2.0×10-12	1.5
unfilled PBD [55]	9.0×10-18	3.5
unfilled PBD [64,65]	7.5×10-16	3.0
filled PBD [55]	4.0×10-14	1.9

Table 4.2. Coefficients ***k*** and exponents ***n*** in unsteady abrasion stage at room temperature [56,57]

Materials	***k***	***n***
unfilled NR	5.9×10-16	2.9
unfilled NBR	(1.0~2.6)×10-23	5.2
filled NBR	(2.1~4.3)×10-15	2.5

Despite the fact that Uchiyama and Ishino [60] used different abrasion testing machine, they also obtained the linear wear-rate equation in the same form as equation(4.8), in which, the corresponding ***k*** and ***n*** is equal to 5.4×10-7 and 1.2

respectively for unfilled IR.

In the unsteady abrasion stage, the variation of the linear wear-rate of rubber is also increased with the sliding distance L in addition to following the equation (4.8) [56,57]. That is:

$$R_L = C\exp(mL) \tag{4.9}$$

Where, the coefficient ***C*** and exponent ***m*** are mainly dependent on the kinds of rubber and magnitude of frictional work, the values of which are shown in Table 4.3.

Table 4.3. Coefficient ***C*** and exponent ***m*** of three kinds of rubber at room temperature [56,57]

Materials	***w*** kJ/m2	***C***	***m***
unfilled NR	0.6~0.7	54.0×10-9	7.7×10-4
unfilled NBR	0.5~0.8	1.8×10-9	5.9×10-4
	0.9~1.0	4.3×10-9	5.7×10-4
	1.1-1.5	96.0×10-9	3.8×10-4
filled NBR	0.6~0.7	1.8×10-9	2.0×10-4

Based on the essential characters of wear, as stated in chapter 3, section 3.1.2, the wear rate in unsteady state is generally lower than that in steady state.

According to the experimental results, Zhang [57] noticed that the unsteady state did not appear again even if the frictional work increased to rather high value, once the steady state was reached under the condition of a smaller work input. This result might be important in practice operation for rubber components as the wear-rate of unsteady state caused by larger frictional work is always higher than that produced by small fractional work based on the equation (4.8). Therefore, a rubber component operated on a heavy frictional work could be developed into its steady state in trial runs by means of a relatively light frictional work before coming into operation under normal conditions. As a result, the loss of materials decreased, and this component might run more smoothly. Obviously, this running-in mode would be helpful to prolong the working life of the machine with rubber components.

The effects of ambient temperature and atmosphere on linear wear-rate were also examined by Gent and Pulford [55]. The partial results are shown in Tables 4.4 and 4.5, where f denotes the frictional coefficient.

Table 4.4. Effect of ambient temperature on linear wear-rates (w=0.6kJ/m2)

Materials	Temperature, ℃	RL, mm/r	*f*
unfilled NR	25	150	1.44
	100	190	1.18
unfilled SBR	50	82	1.58
	100	195	1.37

Table 4.5. Effect of ambient atmosphere on linear wear-rate for filled PB (W=1.6kJ/m2)

Ambient atmosphere	RL, mm/r	*f*
Air	48	1.5
Nitrogen	52	2.2

As seen from the Table 4.1 to Table 4.5, the main factors influenced on the linear wear-rates are mechanical properties, including hardness, tensile strength, tear strength, and fatigue strength and so forth, but the effects of ambient temperature and atmosphere on the linear wear-rates are not so much.

It has been found that the linear wear-rates are reduced obviously while the NBR and PB materials are with the addition of carbon black (see Figure 4.15) [55,57]. The heavier the operating condition is, the better the effectiveness will be. The addition of carbon black does not notably increase the mechanical fatigue resistance of rubber components. Moreover, the unsteady-state rate of wear of filled NBR is less affected by the number of revolutions (sliding distance) than that of unfilled NBR as seen from Table 4.3. The exponent m of unfilled NBR is about three times that of filled NBR under otherwise almost equal conditions. Apparently, these experimental results verify again that the abrasion by small-scale tearing results from not only the mechanical fatigue but the other rupture processes as well [55,56].

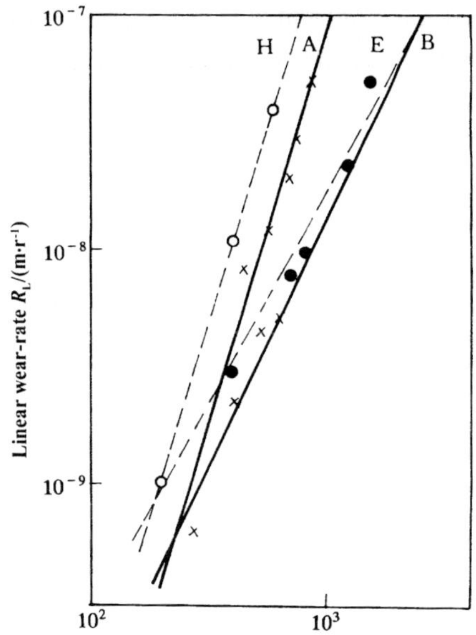

Fig. 4.15. Effect of filler on linear wear-rates [55,57]
A—unfilled NBR; B—filled NBR; H—unfilled PB; E—filled PB

Gent and Pulford observed the reversals in the relative rates of wear of the unfilled PB compared to the unfilled NR and SBR at different severities of wear [55]. From Fig. 4.15, it is also found [56,57] the reversals in the relative rates of wear of the unfilled NBR and PB compared to the filled NBR and PB respectively under different severities of wear. It means that the linear wear-rates of carbon-black-filled rubber are higher than that of unfilled rubber when the frictional work is less than a certain value. This phenomenon probably reflects a competition in different wear mechanisms. It would be deduced that the dominant wear mechanism is mainly micro molecular fracture resulted from the microcutting, when the frictional work is less than a certain value. However, beyond this value, macrodelamination plays a dominant role in the abrasion process, whereas the reinforced effect of carbon black is able to enhance the tensile strength of rubber. Therefore, the linear wear-rates of carbon-black-filled rubber are much lower than that of unfilled rubber under the condition of larger frictional work.

4.1.3 Multiple-Point Contact Abrasion

4.1.3.1 Wear Mechanisms

Generally, the existent dry abrasion is mostly pattern abrasion by multiple-point contact. For this reason a set of hemispheres serves as a model of a concrete surface having sand and gravel particles embedded in cement in early studies. Experiments were conducted respectively with fine and course concrete surfaces as abradants on a single rubber compound at an applied load of 0.16 MPa and a sliding velocity of 0.48 m/sec [18]. The experimental results are given in Table 4.6, where S is the mean spacing between the ridges of the abraded rubber surface.

Table 4.6. Abrasion effects on road surfaces

Type of concrete surface	Coarse	Fine	Coarse/Fine
Rv cm3/m, cm2	1.57×10-3	0.41×10-3	3.83
S cm	0.832	0.234	3.56

These experiments demonstrated [18,19]

$$S \propto R_v \propto d^{\frac{2}{3}} \tag{4.10}$$

Where, d is the mean particle size.

It is considered that this abrasion process is assumed to be due to tensile failure of the elastomers, as shown in Figure4.4, for the needle experiments. Detachment of molecules of the elastomer from the bulk is caused by catastrophic tearing behind the sharp asperities in the track, following very high rates of strain (about 10,000% per second) on rubber surface [13]. As for the origin of ridges, Schallamach [49] considered that it could be interpreted in the same way as the mechanism of abrasion by a point contact. Moreover, the abrasion ridges are resulted from the high ratio between coefficient of friction and dynamic stiffness of rubber. Accordingly, these ridges are more marked on soft mixes than on hard ones, and are also more marked on blunt tracks than on sharp ones, because these ridges can develop fully before they are abraded away.

Since the late 1970's, many Indian scientists have conducted a large number of studies in the multiple-point contact abrasion [67~77]. Especially, Bhowmick et al [67~71] have investigated extensively the wear characteristics and formation process of abrasion pattern on the rubber abraded surfaces by scanning electron microscope (SEM). The ribbed constitution and microstructure were observed respectively on the abraded surfaces of filled NBR or filled NR and unfilled NR materials [68,71]. They considered that the rubber abrasion is caused by two mechanisms as follows. One is the abrasion mechanism resulted from microcutting action and the other is the frictional mechanism arised from the frictional force. The former is followed by the gum vulcanizate and filled NR, but the latter is followed by NBR, SBR and BR [67,71]. Moreover, the abrasion resistance of rubber does not depend on its netted structure [69,70]. In addition, SEM study on the fracture mechanism resulted from tearing of carboxylated nitrile rubber (XNBR) has been made further by Chakraborty and co-workers [73,74].

The abrasion characteristics of thermoplastic polyurthane (TPU), 1.2 polybutadiene

(1,2PB), styrene-isoprene-styrene block copolymer (K1107), plasticized polyvinylchloride (PVC), thermoplastic copolyester elastomer (Hytrel 40D) and a blend of plasticized ployvinylchloride and thermoplastic copolyester elastomer (Hytrel-PVC) have been examined by Kuriakose et al [75] and Thomas [77]. It has been found that the developing process of wear behavior is consistent with the results as stated in Reference [57]. The experimental results demonstrated that the abrasion resistance of Hytrel-PVC blends increased with increasing Hytrel content; moreover, the mechanism of abrasion was also changed with the content. Blends containing higher or equal proportions of Hytrel (≥ 50wt%) showed frictional wear while those containing lower or equal proportions of Hytrel (≤ 25wt%) showed abrasive wear.

As a part of an extensive project dealt with the wear mechanisms of tread rubber, Schweitz and Ahman [78,79] investigated the mild abrasion with low rate of wear of rubber. They simulated the mild abrasion of NBR, SBR and NR-BR against polyester mesh using both pin-disc and pin-cylinder testers. The weight losses were determined by neutron activation analysis. It has been found that the wear behavior and wear rates of rubber are changed with the normal load, sliding velocity and the contaminated extent of its counterface.

An infrared spectroscopic technique was used by Dyrda and co-workers [80] to investigate the worn surface of the unfilled vulcanizate polyisoprene being rubbed against emery cloth. It is noticed that the abrasion is mainly caused by the action of mechanical stress resulting in repeated scission of the polymer chains rather than delamination of the polymer molecules or scission of the intermolecular bonds.

Of recent years, Thavamani and co-workers [81] studied the wear mechanisms of NR, SBR and hydrogenated nitrile rubber (HNBR) vulcanizates under different conditions using SEM. The filled NR and SBR vulcanizates were abraded against granite rock with surface roughness of 5-10 μm. No ridge was observed on the rubber surface at low normal load. However, obvious ridges were found on the worn surfaces after the two kinds of vulcanizates were swollen in toluene. Similarly, the HNBR compound shows typical ridges only if it is swollen in dimethyformamide (DMF). The reason is that the tear resistance and elastic modulus of the swollen rubber are much reduced. Hence, the spacing between ridges observed on swollen rubber surfaces is larger than that on unswollen rubber surfaces. For the three vulcanizates above, the spacing between adjacent ridges increases with the rise of temperature. It would be accounted for that the elastic modulus of rubber is decreased with the increase in temperature.

In addition, Abu-lsa [82] investigated the abrasion characteristics of three elastomer compositions consisting of NR, SBR and EPDM by small particle impacts at a pressure of 276 kPa. A steel shot, a steel grit, and two types of sand particles were used as the impacting particles. The results showed that NR has the highest resistance to abrasion followed by SBR and then EPDM. High temperature, long abrasion time, and high degree of stretch of the rubber sample during abrasion all resulted in a higher degree of abrasion. Moreover, the mechanism of abrasion by particle impacts is initiated by tears generated on the surface of the elastomer, and then propagated by cumulative growth of the cracks. Particle shape and size are also influenced the severity of abrasion.

4.1.3.2 Rates of Wear

(1) Influence of load on ware rates

Experimental results showed that the relationship between linear rates of wear and normal pressure can be regarded as linear as a first approximation for the rubber sliding against sharp-grained counterface (track) (Figure 4.16) [83]. In light of that the frictional force is likewise roughly proportional to the load in this situation [84]. Therefore, we may as a first approximation assume proportionality between linear wear-rates of rubber abrasion and frictional force. According to the mechanism of rubber abrasion by a point contact as stated previously, it is thought that the energy density at break, a relevant strength characteristic of the rubber, is the basis of the above relationship. Thus, the depth of abrasion per unit slip path, namely linear wear-rate RL is given by the following expression [49]:

$$R_L = \frac{kfp}{U} \tag{4.11}$$

Where, k—constant,

p—normal pressure,

U—energy density at break,

f—coefficient of friction.

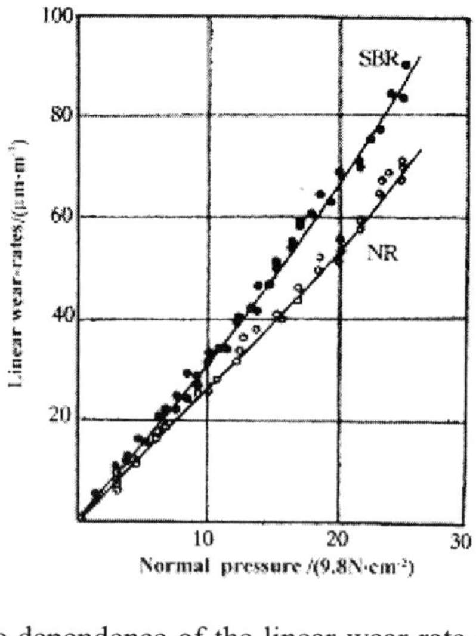

Fig.4.16. Normal pressure dependence of the linear wear-rate of SBR and NR materials sliding on emery cloth [83]

Grosch and Schallamach [85] pointed out that the equation above is only applicable for the intrinsic abrasion.

Under the condition of blunt-grained counterfaces (tracks), such as metal gauge, concrete road or severely worn grinding wheel, the particles are detached from the rubber surface by repeated attack of the blunt asperities on the counterfaces. Experimental results demonstrated that the relationship between linear wear-rates and load on such situation is no longer existed as proportionality but a power law as follows [49,86]:

$$R_L = (p / \rho_o)^n \tag{4.12}$$

Where ρ_o and n are empirical values.

Experimental values of n are given in Table 4.7

Table 4.7. Values of the exponent n for four types of tracks [86]

Elastomers	Grinding Wheel A	Concrete Floor B	Concrete Road C	Tar/macadam Road D
NR	1.37	1.49	1.31	1.57
SBR	1.45	1.52	1.57	1.96
BR	1.21	1.30	2.23	2.25
NBR	1.51	1.53	1.88	1.90

The curves in Figure 4.17 show the normal pressure dependence of the linear wear-rates of a NR tread rubber on four kinds of blunt-grained tracks A, B, C, D as listed in Table 4.7.

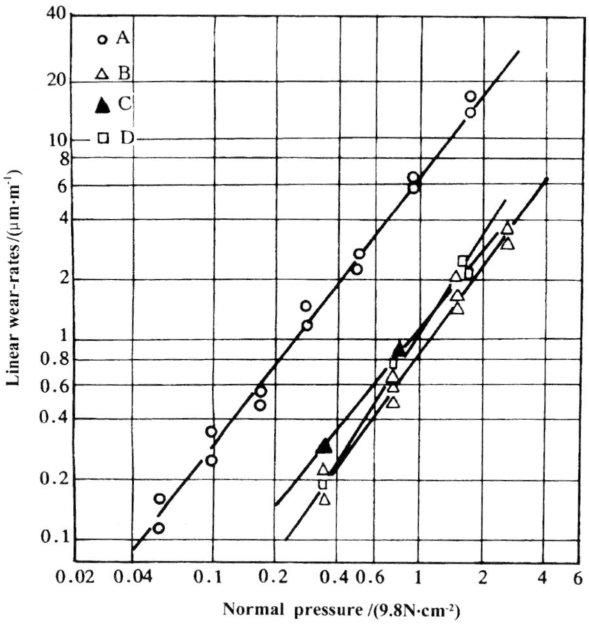

Fig. 4.17. Normal pressure dependence of the linear wear-rates of a NR tread mix on various blunt tracks [86].

For the case of an elastomer sliding on a rough textured surface, the volumetric wear-rate of rubber abrasion R_v (cm3/cm2, m) is related to the interfacial pressure p by a power law as follows [87]:

$$Rv = Cp^{\alpha} \tag{4.13}$$

Where C and are constants depending on the properties of rubber and the texture of the abrading surface as well as the types of wear.

For intrinsic abrasion, =1, namely, the relationship between volumetric wear-rate and normal pressure is a line through the origin of coordinates (Figure 4.18) [85].

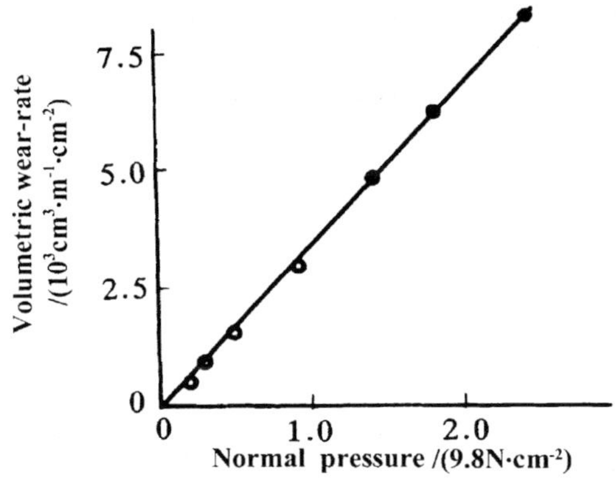

Fig. 4.18. Normal pressure dependence of the volumetric wear-rates of intrinsic abrasion of NR [85]

The volumetric wear-rates of rubber on blunt tracks are much sensitive to the antioxidant and atmosphere (oxygen and nitrogen).

(2) Influences of velocity and temperature on wear rates

The rates of wear of intrinsic abrasion are affected strongly by the sliding velocity and temperature. In general, the dependence of temperature on velocity cannot be obtained according to the experimental original relationship between the abradibility and velocity at various temperatures. Thus, it should be transformed into the set of corresponding master curves, which means an equivalence of the temperature- and rate-dependence of the abrasion. Hence, following the time-temperature equivalent principle of polymer and the WLF Equation formulated by Williams, Landel and Ferry, it can be set up a velocity-temperature equivalent principle, namely, this equivalence is realized by introducing a displacement factor α_T . According to experience, each polymer has a characteristic temperature T_s, lying about 50K above the glass transition temperature and with which as reference temperature the shifts α_T leading to the master curve obey the following applicable equation [49]:

$$\lg a_T = -8.86(T - T_s) / [101.6 + (T - T_s)] \tag{4.14}$$

The changes of abradibility of intrinsic abrasion with velocity for SBR and NBR are shown in Figure 4.19. These master curves are transformed according to a reference temperature of 20℃.Another master curves for the velocity dependence of the abradibility of three unfilled rubbers are illustrated in Figure 4.20 [85]. A similar dependence is obtained for reinforced and strain-crystallising elastomers, although the velocity-temperature equivalence principle is no longer valid in these cases [88]

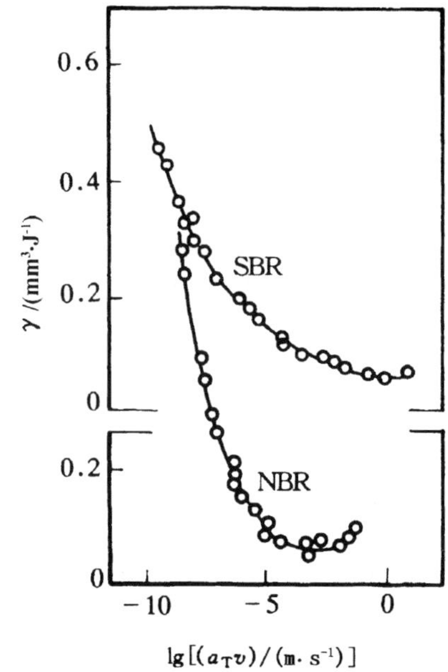

Fig.4.19. Velocity dependence of the abradibility of SBR and NBR [85,88]

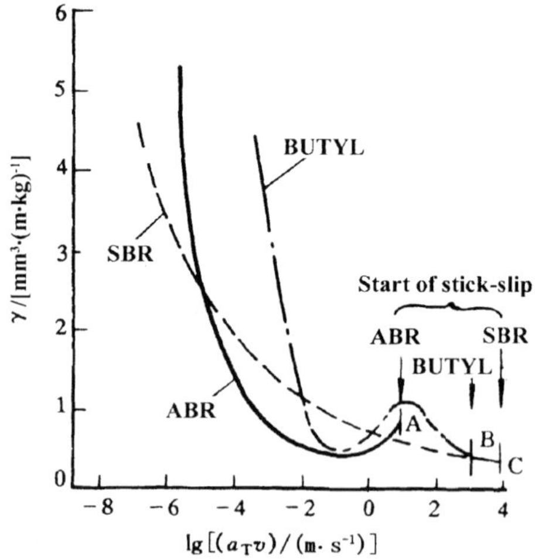

Fig.4.20. Velocity dependence of the abradibility of three unfilled rubbers [13,85]

It is seen from Figures 4.19 and 4.20 that all of the curves have one or two minima of abradibility γ. A qualitative explanation for this can be obtained from Fig. 4.19, which graphically depicts the variation of f, λ, γ with the logarithm of v at given operating temperature [13]. It is known from equation (3-5), $\lambda \propto \frac{1}{G}$, then

$$\gamma \propto (f, G)^{-1} \tag{4.15}$$

According to the friction mechanism of elastomers [19], we obtained:

$$f_{Tot} = f_A + f_H \tag{4.16}$$

Where , f_A —coefficient of adhesion friction

f_H —coefficient of hysteresis friction

Thus, the curve as shown in Figure 4.21(a) can be obtained by superimposing the separate contributions of adhesion and hysteresis to the frictional coefficient. Owing to the viscosity of elastomer, the variation of shear modulus of elastomer G with lg$v$ at constant temperature may be sketched as in Figure 4.21(b). The variation in the abrasion factor λ is also shown in this figure. Moreover, the curve in Figure 4.21(c) can be obtained from Figure 4.21(a) and (b) in accordance with equation (3.7) in *Chapter 3*. Obviously, each minimum of the curve in Fig.4.21 (c) corresponds approximately to a maximum in the friction-velocity curve (Fig.4.21 (a)). Therefore, an adhesion minimum C and a hysteresis minimum D for the abradibility γ are observed in fact [13].

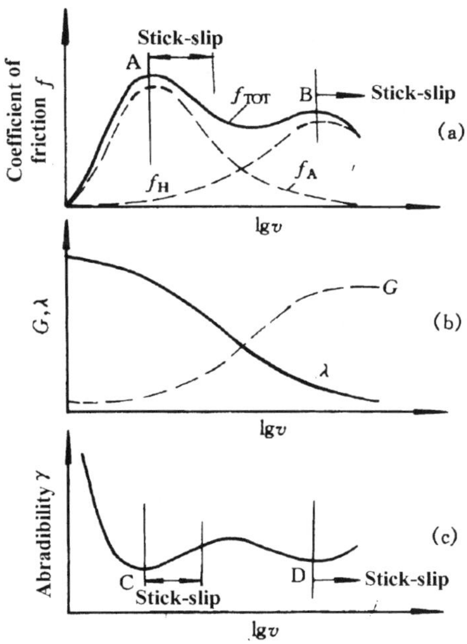

Fig.4.21. The viscoelastic nature of rubber abrasion [13]

Figure 4.22 shows the temperature dependence of γ, λ and f at a given sliding velocity of 1cm/s for the case of four different unfilled non-crystallising rubbers [13,85]. The varying tendency of these curves is in consistent with that of those curves as shown in Figure 4.20. It might be concluded that rubber abrasion has distinct viscoelastic properties and that it generally can be predicted from frictional data in accordance with equation (3.7) [19]. The stick-slip region between the parallel vertical lines lies in the temperature range in which the coefficient of friction has a maximum.

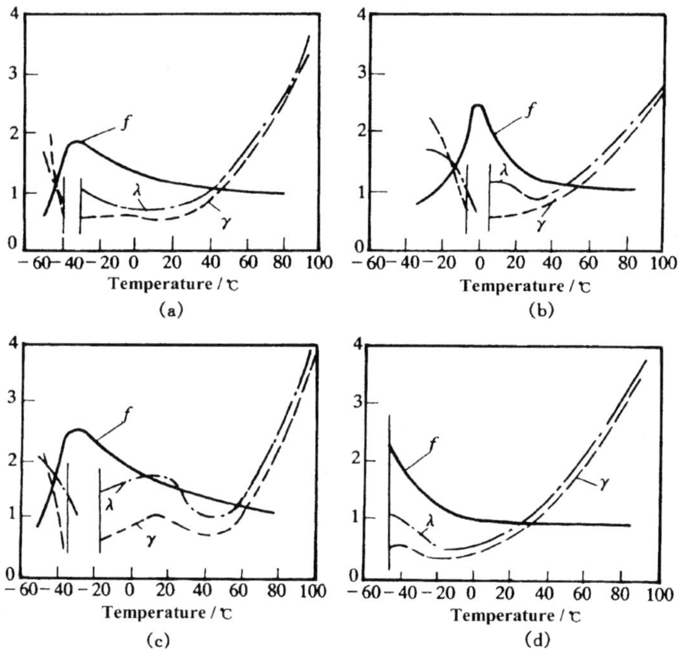

Fig.4.22. Temperature dependence of friction and abrasion for non-crystallizing unfilled rubbers [13, 85]

(a) SBR; (b) ABR; (c) Butyl; (d) NR

(3) Influence of antioxidant and atmosphere on wear rates

Schallamach carried out a series of experiments on the effects of antioxidants and ambient atmosphere on the abrasion of rubber mix using an Akron abrader [89]. These experiments were conducted on a normal aluminium oxide grinding wheel which had been blunted by fairly long service. A dust mixture of two parts silicon carbide and one part fuller's earth was fed in at intervals between specimen (rubber wheel) and grinding wheel, to counter smearing as far as possible. Figure 4.23 shows the variation of wear rates of NR materials protected and not protected by antioxidant with frictional distance (revolutions of grinding wheel), when the atmosphere is changed from air to nitrogen and then to air again.

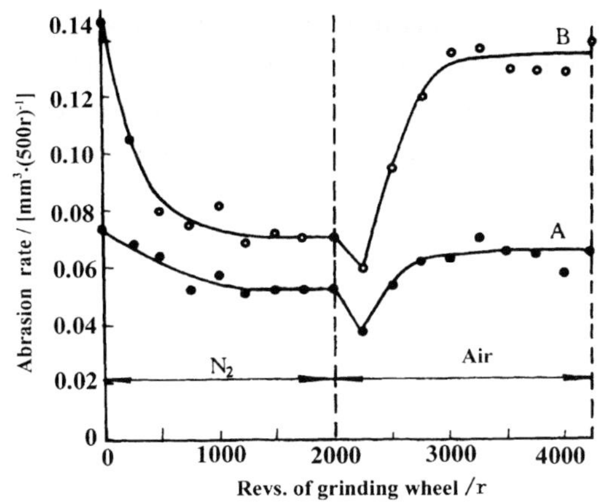

Fig.4.23. Effects of antioxidant and atmosphere on abrasion rate of NR tread mixes [49,89]
A—protected by antioxidant (Nonox ZA); B—no antioxidant

As seen from Figure 4.23, the wear rates of the unprotected NR mix in air are almost twice that of the mix protected two parts of Nonox ZA. When the measurements are continued in nitrogen, the abrasion rates fall, but an equilibrium state sets in only after about 2000 revolutions of the grinding wheel. The relative difference in wear rates between the two mixes is then much less than in air. In the subsequent measurements in air the first reading in either case is still lower than the equilibrium value in nitrogen, and the abrasion rate only then rises slowly again to its original level [89]. Brodskii and co-workers [90] also reached the similar conclusions according to the experiments by using two different slide tracks corundum paper and a knurled steel plate as the counterfaces of rubber.

The data in Table 4.8 show the equilibrium values of wear rates of three types of rubbers under the various test conditions using Akron abrader [89]. The dust mixture listed in Table 4.8 is mixed by two parts silicon carbide with one part fuller's earth. As seen, for these two antioxidants, the wear-resistance of Nonox ZA is better, though the rubber abrasion is affected by the counterface (grinding wheel) and dust mixture in this case. The smearing resulted from the oxidative degradation of the rubber surface during the abrasion naturally reduces the abrasion [89]. It is known that magnesium oxide adsorbs contamination in the contact surface more than fuller's earth; on the other hand, fuller's earth, on account of its larger particles, is a more effective abradant [85].

Uchiyama investigated the effect of environment on the abrasion of both NR and SBR vulcanizates contained antioxidant functional groups [91]. Experiments were performed under various environmental conditions, including in vacuum, in air oxygen, nitrogen and argon. A rubber pin was loaded against an abrasive cloth, emery paper or metal gauze respectively. The experimental results showed that oxygen increased the wear rates of both NR and SBR vulcanizates under the conditions of various environmental atmospheres. However, the effect of antioxidant on the rubber abrasion is depended on the rubbing condition. On the mild frictional condition, the abrasion is reduced markedly by the action of antioxidant.

Table 4.8. Wear rates of rubber (mm3/500 revs. of Akron wheel) under various conditions (excerption) [89]

Rubber		NR		SBR	BR
Antioxidant	Nonox ZA 2	Nonox HFN 2	none	Nonox HFN 2	Honox HFN 1
Grinding wheel, dust mixture					
Air	32.9	42.5	66.5	25.2	18.1
N2	26.1	30.6	36.4	27.0	25.2
Aluminium wheel, dust mixture	--				
Air	--				--
N2		8.0	12.2	9.6	--
		14.8	15.8	13.2	
Aluminium wheel, Mgo					
Air	15.5	25.4	40.2	10.7	0.6
N2	4.0	5.2	5.9	2.0	3.7
Stainless steel wheel, dust mixture					
Air	23.4	30.8	48.5	14.8	4.5
N2	18.7	23.0	26.8	20.9	19.7
Stainless steel wheel, Mgo					
Air	38.8	52.7	75.0	25.4	12.2
N2	10.4	11.9	12.2	7.8	5.0

4.2 Oily Abrasion

4.2.1 Basic Features

In the late 1960's and the early 1970's, Schallamach [89], and Boonstra et al [92] respectively described a wear phenomenon, in which a typical tire tread material was transformed under relatively mild dry abrasion conditions into a sticky layer on the rubber surface and produced oil debris. This type of abrasion is denoted as oily abrasion by the present author [56]. It is also called smearing or clogging in some references.

In the wear process of oily abrasion, no dry debris but only a less smooth and shiny particulate debris are produced and clogged on the worn surface. The wear phenomenon of a gum NR vulcanizate was investigated by the present author [56] under the condition of mild abrasion in unsteady state. It has been observed on the worn surface that the sticky layer is also formed in the shape of ridge pattern. Examination of this ridge pattern with an optical microscope revealed clearly that the appearance and arrangement of ridges are very uniform and the peak of ridges has rounded cross section (Figure 4.24) [56].

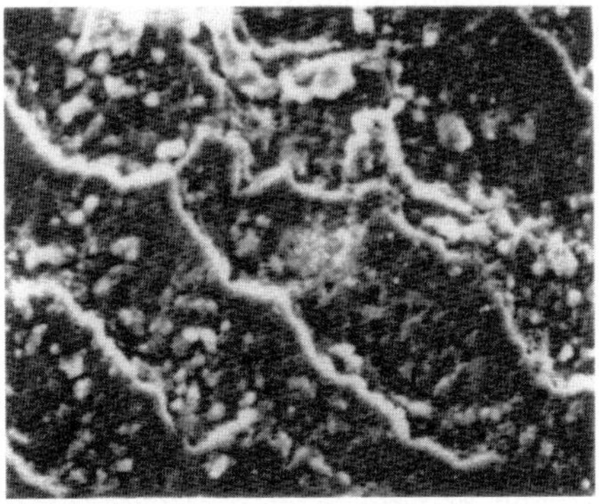

Fig. 4.24. Oily-abrasion pattern of NR ($\overline{F} = 0.41 kN / m$)

4.2.2 Wear Mechanisms

Although it is not disputed that the primary mechanism of oily abrasion generated under mild abrasion condition for certain rubber compounds, such as carbon-black-filled compounds of SBR, NR and ethylene propylene rubber (EPR) is due to rubber degradation, views are divergent on what cause is responsible for the decomposition process. A more prevalent viewpoint is that chemical degradation due to the thermal effect [89] and the chemical bonds of rubber being attacked by light and oxygen [93]. Moreover, the operating conditions of heavy load and high speed can raise the temperature of frictional surface and as a result accelerate the degradation process. However, antioxidant is considered to reduce the amount of wear [93].

Later on, the chemical effects being responsible for the oily abrasion of various elastomers were studied further and extensively [55,56,88,94~99]. Three views on what cause is accounted for the chemical decomposition process were put forward [55]:

(1) Thermal decomposition due to local heating during sliding;

(2) Oxidative degradation possibly accelerated by local frictional heating;

(3) Mechanochemical decomposition initiated by mechanical rupture of macromolecules to form reactive radical species;

The first view is not quite correct since certain elastomers are also sensitive to thermal decomposition but no degradation during sliding. As for the second view, it is also not the principal cause of decomposition as some rubber compounds even in an inert atmosphere are also decomposed by friction [95]. The last standpoint is essential for degradation during sliding because the extent of decomposition of the rubber in air and in an inert atmosphere respectively is closely related to the variation of molecular weight of non-cross-linked rubber under the action of continuous shearing [95,96]. However, a few possible effects of frictional heating cannot also be ruled out completely.

Based on the experimental results of the wear process of gum NR vulcanizate in unsteady state under mild abrasion conditions ($\overline{F} = 0.24 \sim 0.41 kN / m$) [56], it is considered that under the action of microcutting by the frictional force, a number of macromolecules of the rubber rupture, namely fracture of molecular chains. As a result, the reactive free-radicals form at the end of fracture. These radicals react with atmospheric oxygen or interact with each other depending on their intrinsic reactivity and environment [55]. Hence, the mechanochemical decomposition is initiated, and then a sticky layer forms on the frictional surface. Moreover, this layer produces viscous deformation and takes the shape of a number of parallel ridges under the action of the frictional force. However, the microstructure of worn surface is not layered or fibered, but honeycombed, as shown in Fig. 4.24. The spacing of the ridges is generally larger than that of the dry abrasion pattern under otherwise equal conditions, and is also increased with the increase in frictional force.

In general no tear and tensile failure will occur on the rubber surface due to the sticky layer formed on this surface. Moreover, the wear rates decrease or even fall to zero as the rubber decomposition proceeds. Obviously, the sticky layer or viscous liquid-like film retained on the rubber surface acts as an protective layer or lubrication film, which can also protect the rubber surface to some degree from further degradation. Gent [97] has pointed out yet that the properties (viscosity and stickability) of the sticky layer are dependent on the reaction induced by the mechanical fracture of macromolecular and on the particular composition used in the rubber formulation, especially, that composition being able to join the free-radical reactions.

Muhr and Thomas [98] have investigated the oily abrasion in the case of no any large-scale mechanical abrasion. It has been found that the wear rates of rubber are closely related to its cross-linking property. The weaker the cross-linking, the higher the wear rate is. Possibly, it is accounted for that the cross-linking is easily destroyed by the mechanical stress. Therefore, they considered the oxidation activated by the mechanical stress instead of the thermooxidizing induced by the frictional heat to be possibly the main cause to generate this type of abrasion.

The antioxidant may lower wear rates of oily abrasion as mentioned above. Thus, the effect of antioxidant, which was studied long ago, is again riveted attention several years ago [99].

Muhr and Roberts [100] investigated the influence of ambient temperature on oily abrasion using a peroxide-cured NR wheel (Akron abrasion type) sliding on a smooth glass plate. The testing ambient temperatures are in the range of $-30{\sim}60\,^{\circ}c$. Both smear and rubber particles were deposited on the glass plate after rubber wearing. The higher the ambient temperature, the greater the production of smear was, and the lower the temperature, the greater was the amount of particulate debris left on the plate with less smear. Below $-20\,^{\circ}c$ there appeared to be no smear, only copious particle production. It is worthy of note that they conducted an experiment, in which the rubber wheel was run against smooth-surfaced ice at $-28\,^{\circ}c$. In this situation, the ice is practically acted as an interface temperature limiter ($0\,^{\circ}c$ maximum). In the event, only particulate rubber debris was obtained but no smear, it would appear that interface temperature greater than $0\,^{\circ}c$ is required for smearing.

4.3 Wet Abrasion

As an independent kind of wear, wet abrasion or hydro-abrasion has not been investigated extensively yet. However, many rubber-metallic tribo-components used in a great number of fluid machinery are operated in abrasive mud, oil or water media containing or without containing solid particles. Obviously the wet abrasion is usually the dominant cause resulting in earlier failure of these components.

4.3.1 Point-Contact Wet Abrasion

Late in 1980's, the physical process of point-contact wet abrasion of NBR materials was investigated by the present author using a modified Tabor-Eldredge friction-testing machine. Experiments were conducted under the condition of lubrication with polydimethylsiloxane varied in viscosity (9.4mPa·s, 94 mPa·s and 950 mPa·s respectively). It has been shown that under the conditions of identical testing parameters and method (refer to chapter 2, section 2.3.2.), its wear mechanism is all the same to that of dry abrasion, but the extent of tearing is mild (Figure 4.25).

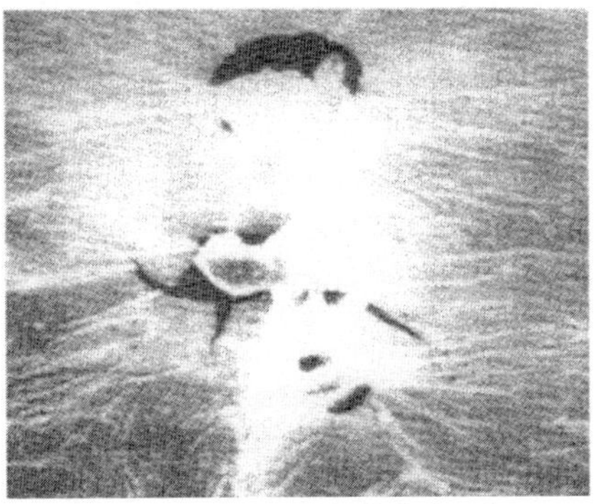

Fig. 4.25. Tearing on the surface of NBR (lubrication, cone apex angle (10°)

Later on, using a modified pin-disc tester, the physical process of wet abrasion by a point contact in the medium of water was examined with a metal needle of 2.5mm in diameter. The needlepoint with a cone apex angle of 30° is sliding against a rotating nitrile rubber disc. Experiments were conducted at a sliding speed of 0.33m/s and a normal load of 1.96N. The plough grooves were observed on the rubber surface obviously and the tearing layered structure appeared at the base of groove (Figure 4.26), moreover, some debris occurred (Figure 4.27).

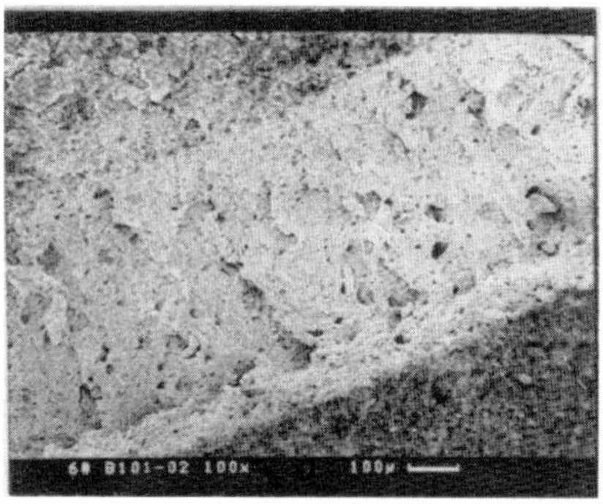

Fig. 4.26. Worn surface of NBR

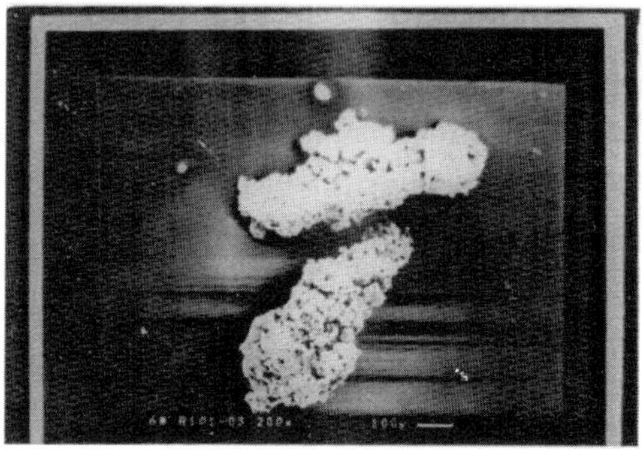

Fig. 4.27. Wear debris

4.3.2 Line-Contact Wet Abrasion

Muhr and Thomas [98] examined the effect of water as a lubricant on the wear rate of wet abrasion by a line contact using a blade abrader. It has been found that as compared with the wear rate of dry abrasion, the wear rate of wet abrasion is dramatically reduced by factor of ten at least, whereas the frictional force decreases only by perhaps 10%. This result is in conflict with the theory proposed previously by Southern and Thomas [54]. Other liquid lubricants such as silicone oils also show a similar effect. In order to check if the above results were to an extent fortuitous, experiments were carried out with a blade making a trailing angle of 45° with the surface of NBR materials under the condition of dry and wet respectively. Furthermore, the experimental results are compared with the calculating values based on the theory stated in Reference [54], which are shown in Figure 4.28. Whether the blade made a trailing angle or not, the experimental results under dry condition show fair agreement with the theoretical values, but the discrepancy is obvious under water lubricated condition.

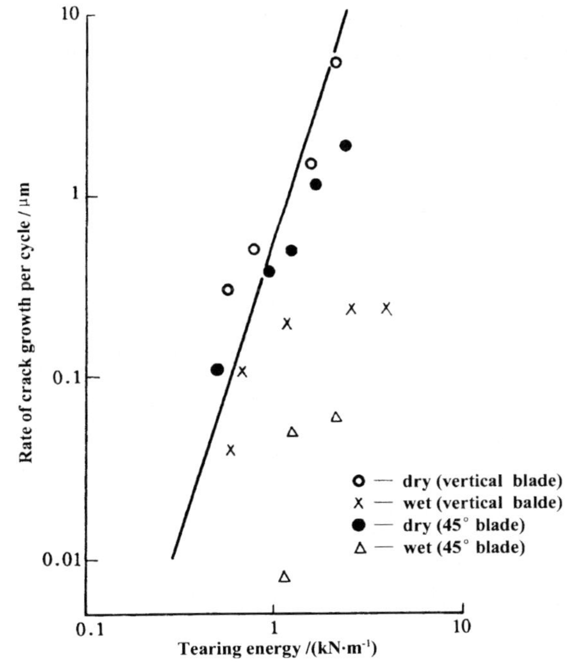

Fig.4.28. Effect of lubrication and blade angle on wear rate for SBR gum rubber [98]

A possible reason for the discrepancy is suggested by their observation [98] that under lubricated conditions the abrasion pattern and its formation process are different from those under dry conditions, namely the tongues without being flipped over and being merely compressed as indicated in Figure 4.29. According to this physical model, the tearing energy may be approximately calculated for a tongue of thickness a as

$$T = (F / b)(1 - \lambda) - Ua \tag{4.17}$$

Where, F —frictional force,

b —width of the rubber wheel,

λ —extension ratio of the tongue (less than unity),

U —stored elastic energy per unit volume in the compressed tongue.

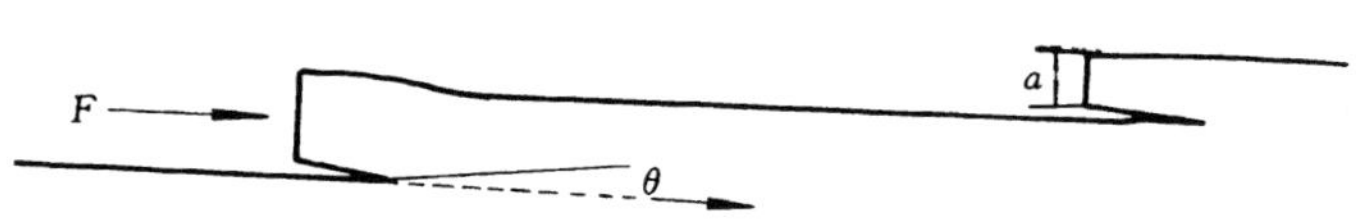

Fig 4.29. Deformation of pattern under lubricated conditions [98]

As seen from the equation above, a change of this sort in the mode of deformation of the pattern under lubricated conditions is the source of the large reduction in wear rate.

4.3.3 Multiple-Contact Wet Abrasion

According to that whether or not the site of the solid particles in liquid medium is fixed, there are two kinds of multiple-contact wet abrasion. One is a wet abrasion acted by free particles, and the other wet abrasion is acted by fixed particles, which is generally denoted a wet abrasion resulted from the roughened hard surface sliding against the rubber surface in water medium.

4.3.3.1 Wet Abrasion Acted by Free Particles (Three-Body Abrasion)

Of recent years, Zhang et al [101,102] have investigated the physical process of wet abrasion for NBR material (shore-hardness 65) using a pin-disc wear test machine (Figure 4.30). Experiments were carried out with a rubber pin (6mm OD and 12mm length) sliding against a disc made of steel 45 and immersed in the drilling liquid with different sand contents. The abraded surfaces of samples were examined using scanning electron microscopy (SEM) and X-ray photoelectron spectrometer (XPS).

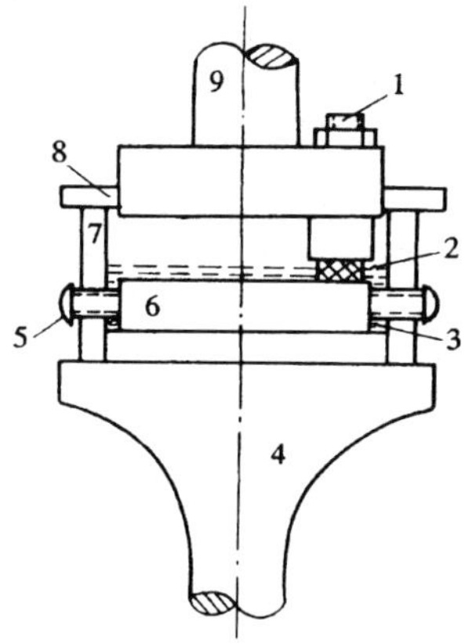

Fig.4.30. Schematic drawing of pin-disc-type test machine
1—specimen holder; 2—specimen(pin); 3—abrasive liquid;
4—rotating shaft; 5—catch bolts; 6—steel disc;
7—container; 8—seal ring; 9—stationary shaft

Experiments were conducted with two sets of test parameters respectively as listed in Table 4.9.

Table 4.9. Test parameters

Sets	No.1	No.2
Number of pin	1	2
Normal load N, N	70~200	44~59
Sliding speed υ , m/s	0.33	1.1~1.8
Sand contents, wt%	2.81, 5.11	0.02, 0.13
Temperature, $^\circ C$	26 ± 3	25 ± 3

(1) Physical process of abrasion

The basic morphological character of worn surface was found to be a number of parallel tearing traces with uneven spacing and microlayered surface texture (Figure 4.31). The spacing of tearing traces decreases roughly with increasing sand content in liquid under otherwise identical conditions (Figure 4.32), and the depth of traces increases with increase in normal load (Figure 4.33).

Fig. 4.31. Three-body wet-abrasion pattern (Set No.1, sand contents 2.81 wt%, normal load 120 N)

Fig. 4.32. Three-body wet-abrasion pattern (Set No.1 Sand contents 5.11 wt%, normal load 120N)

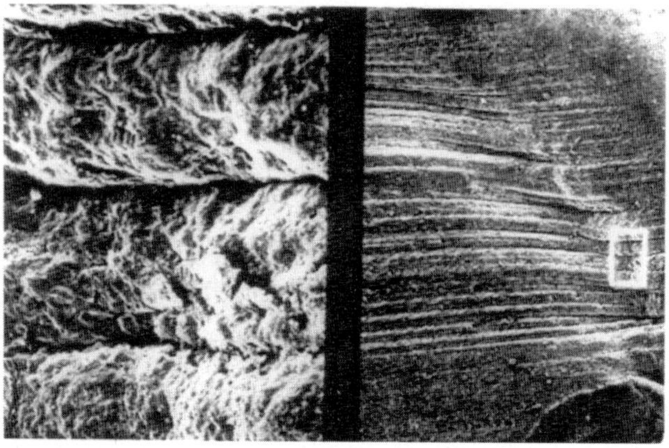

Fig. 4.33. Three-body wet-abrasion pattern (Set No.1, Sand contents 5.11%, normal load 180N)

For the sake of contrast, the morphology of worn surface of wet abrasion in water (i.e., two-body wet abrasion acted by fixed particles) under otherwise identical conditions is shown in Figure 4.34. As seen, some parallel scratching traces with uneven spacing and microlayered surface texture are also observed on the worn surface.

Fig. 4.34. Two-body wet-abrasion pattern (Set No1, normal load 120N)

However, the edges of the tearing traces resulted from the three-body wet abrasion are less regular than that of the scratching traces as illustrated in Figure 4.34. Thus, it might be deduced that the ploughing action on the rubber surface acted by the irregular-shape free abrasive particles along with the direction of motion is ascribed to not only sliding but rotating friction as well.

In general, the growth direction of abrasion pattern of microlayered structure on the worn surface is almost at right angle to the direction of motion (Figures 4.31 and 4.32). However, while the normal load increases by a limiting value, the growth direction of abrasion pattern is turned in parallel with the sliding direction (Figure 33). It would be accounted for that the flowability of sands in liquid is much worsened under heavy load contact, thus the rubber specimen seems to be worn against a filament gauze. In this situation, an abrasion pattern, which is just like the wear pattern formed for NBR material sliding against a polyester gauze [78], is generated on the microlayers along with the sliding direction.

The fineness of the microlayered surface texture indicates the abrasion level of worn surface. The micro-delaminating mechanism probably results from the micromolecular fracture or repeated rupture of molecular chains under the action of mechanical stress as pointed out previously in references [56, 57, 80]. However, this deduction remains to confirm by experimental study.

Consequently, it could be considered as that the mechanism of three-body wet abrasion is involved two kinds of physical process simultaneously, namely, directional micro-tearing and directionless micro-delaminating or polishing under the action of coarse particles and fine ones respectively.

(2) Rates of wear

According to the experimental results of wet abrasion of NBR materials as illustrated is Figure 4.35, it is indicated that the wear rates increase with the frictional force exponentially, which can be represented by a general expression:

$$A = kF^n \tag{4.18}$$

Where, A—weight losses per cycle, g/r;

F—frictional force, N;

k, n -coefficient and exponent. Their values are dependent on the properties of examined materials and the sand content of liquid medium (Table 4.10).

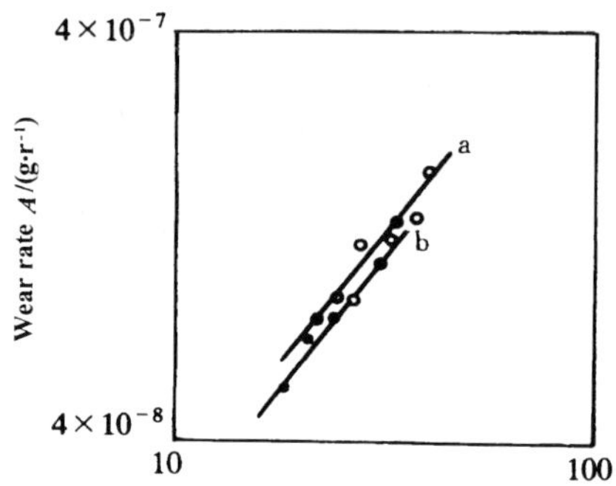

Fig.4.35. Wear rates A plotted against frictional force F (According to the test parameters of first set)

a—Sand content 5.11wt%; b—Sand content 2.81wt%

Table 4.10. Coefficient k and exponent n for NBR at room temperature (Based on the test parameters of first set)

Sand content, wt%	k	n
5.11	2.0×10-9	1.2
2.81	1.0 ×10-9	1.4

It is worthy of note that the wear-rate equation in question has the same form as that obtained previously using a different type test machine under the condition of dry abrasion [57]. The probability is that the wear process of three-body wet abrasion is similar to that of two-body dry abrasion.

Based on the experimental results, the relationship between the wear rates and the coefficients of friction was obtained as shown in Figure 4.36; it can be expressed as following equation:

$$A = af^{-b} \tag{4.19}$$

Where, the values of coefficient a and exponent b are given in Table 4.11.

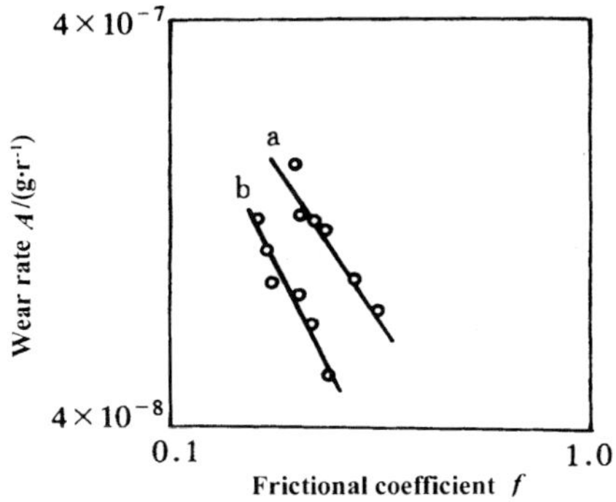

Fig.4.36. Wear rates plotted against frictional coefficient (according to the experimental parameters of first set)

a—Sand content 5.11wt%; *b*—Sand content 2.81wt%

Table 4.11. Coefficient *a* and exponent *b* for NBR at room temperature (based on the test parameters of second set)

Sand content, wt%	*a*	*b*
5.11	0.16	1.5
2.81	0.03	2.0
drilling fluid [103]	0.0034~0.0045	0.025

The above correlation of the wear rates and frictional coefficient is similar to that obtained by Burr and Marshek [103] using different type of test machine and experimental method. As seen from this equation, the wear rates depend on the frictional coefficient to a great extent. However, the coefficient of friction is increased with the decrease in normal road (Figure 4.37), namely

$$f = cN^{-d} \tag{4.20}$$

Where the values of c and d are given in Table 4.12.

The above equation is in agreement with the empirical relation of polymeric friction at low load condition proposed by Tabor [12].

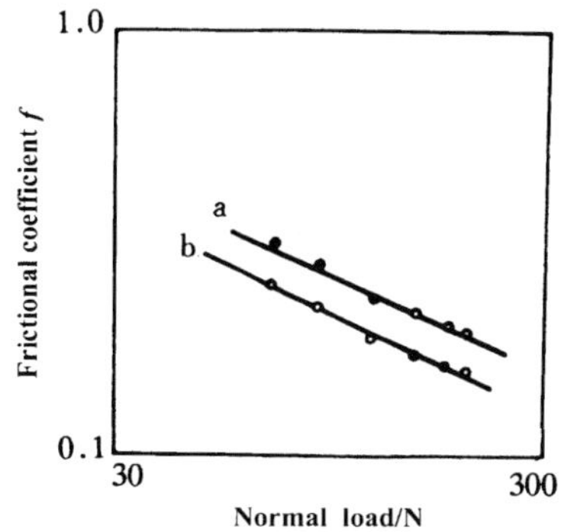

Fig. 4.37. Coefficient of friction f plotted against normal load N (based on the test parameters of the first set)

a—sand content 5.11 wt%; b—sand content 2.81 wt%

Table 4.12. Coefficient c and exponent d for NBR at room temperature (based on the test parameters of the first set)

Sand contest, wt%	c	d
5.11	0.27	0.45
2.81	0.22	0.47

The wear rates and frictional coefficient under the condition of smaller normal load, namely the test parameters of the second set are listed in Tables 4.13 and 4.14 respectively.

As seen from the Tables 4.13 and 4.14, both the wear rates and the coefficient of friction are increased with the decrease in sliding speed. It is probably an improvement on the local lubrication behavior of the frictional surface due to the raise of sliding speed. However, it must be pointed out that the friction mechanism of rubber under wet abrasion condition is still remained obscure.

Table 4.13. Wear rates of NBR (based on the test parameters of the second set) (g/r)

Sand content	Sliding speed	Normal load, N		
wt%	m/s	44	51.5	58.9
0.13	1.1	11.0×10-9	13.3×10-9	21.4×10-9
	1.8	4.17×10-9	6.20×10-9	7.24×10-9
0.02	1.1	1.0×10-9	0.95×10-9	0.60×10-9
	1.8	0.88×10-9	0.87×10-9	0.44×10-9

Table 4.14. Coefficient of friction for NBR (based on the test parameters of the second set)

Sand content wt%	Sliding speed m/s	Normal load, N 4.4	51.5	58.9
0.13	1.1	0.31	0.33	0.33
	1.8	0.14	0.19	0.21
0.02	1.1	0.19	0.16	0.15
	1.8	0.13	0.12	0.11

In addition, the hardness of rubber and the properties of medium have a considerable influence on the wear rate of rubber.

We have investigated the wet-abrasion behaviors of NBR in different drilling muds containing iron ore-stone powder and baritic powder respectively, using a modified mild-wear test machine (Figure 4.38). In experiments, a steel pin of 5mm in diameter was sliding against a rubber disc (Φ ˜ mm×Φ12.0mm) with shore-hardness of 55 and 85 respectively, which are correspondingly termed soft and hard rubber in the following. The steel disc was oil hardened and tempered to a hardness of 55RC, and then chromizing. The shaft was loaded in the range 2.94N to 14.7N and rotated ranging from 100 rpm to 500 rpm. Experimental results are listed in Table 4.15.

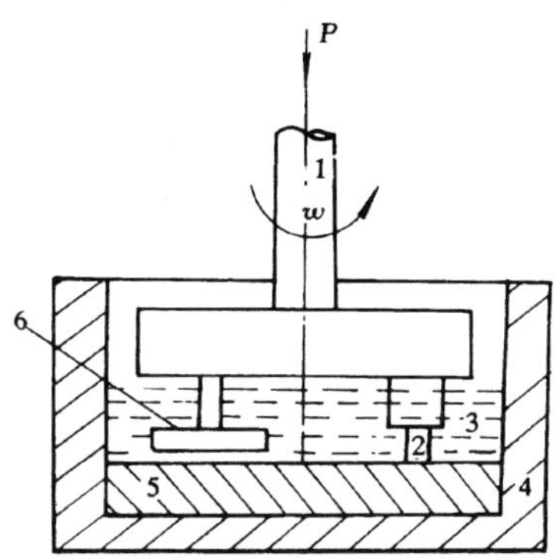

Fig. 4.38. Sketch of mild-wear test machine

1—shaft; 2—cylindrical pin; 3—liquid meduim;4—container; 5—disc; 6—mixer

Table 4.15. Average values of wear-rates of rubber samples (mg)

Liquid media	Mud containing iron-stone powder	Mud containing baritic powder
Soft rubber	10.186	35.409
Hard rubber	5.270	10.814

As seen from Table 4.15, the wearability of the soft rubber is much lower than that of the hard rubber. Moreover, the wear value in mud containing iron-stone powder is much smaller than that in mud containing baritic powder, which might be ascribed to the latter mud being acidulous since the acidoresistance of NBR materials is rather poor. Therefore, despite the hardness of the iron ore-stone is larger than that of the barite so as to tear and scratch the rubber more severely, but the wearability of rubber in the mud containing iron ore-stone powder is still higher. It might be accounted for the combined effect of the chemical and mechanical actions.

4.3.3.2 Wet Abrasion Acted by Fixed Particles (Two-Body Abrasion)

We have also conducted a study of wet abrasion acted by fixed particles using an abrasion tester as shown in Figure 4.39, in which the abrasive paper fixed on the rotating shaft is sliding against the rubber specimen made of NBR materials in water medium. In experiments, the normal pressure (special pressure) on the contact surface of pin was 0.177, 0.354, 0.531 and 0.708 MPa respectively. The abrasive paper (counterface) was rotated and resulted in a sliding speed of 0.335, 0.502, 0.670 and 0.837 m/s respectively in the wear track.

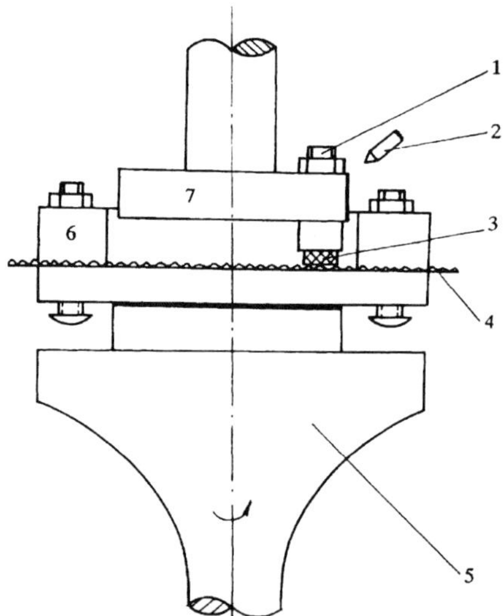

Figure 4.39. Schematic drawing of fixed particle abrasion tester
1 – specimen holder; 2 – nozzle; 3 – specimen; 4 – abrasive paper;
5 – rotating shaft; 6 – abrasive paper holder; 7 – stationary axis

(1) Physical process of wear

A number of parallel and tidy tearing grooves and deformation of the grooves' edge were observed on the abraded surface (Figure 4.40). Thus it can be seen that under the action of fixed particles, the essential physical process of wet abrasion of the NBR material is still micro-tearing resulted from the ploughing by particles. However, in comparison to the wet abrasion acted by free particles, the number of the tearing grooves on the abraded surface is increased and their arrangement is much tidy and even since the number of particles is increased and their sites are fixed.

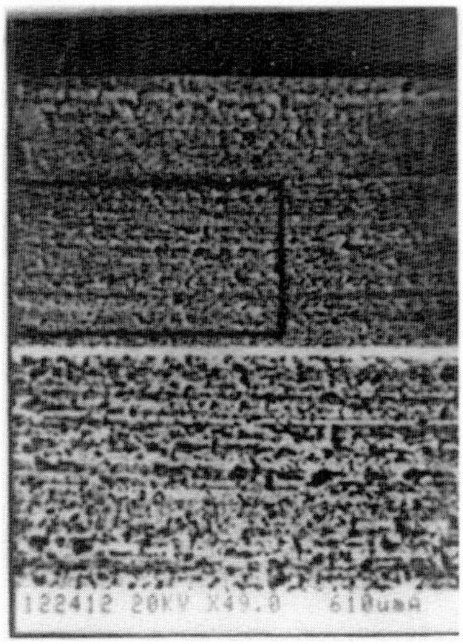

Fig. 4.40. Abrasion pattern of NBR ($p = 0.177MPa$, $v = 0.670m/s$, particle size 63 μm)

(2) Rates of wear

The main influence factors on the wear rate A are generally specific pressure p, sliding velocity υ and particle size δ. Along with the increase in specific pressure, the depth indented into the rubber surface by fixed particles is also increased. As a result, the wear rate rises exponentially (Figure 4.41).

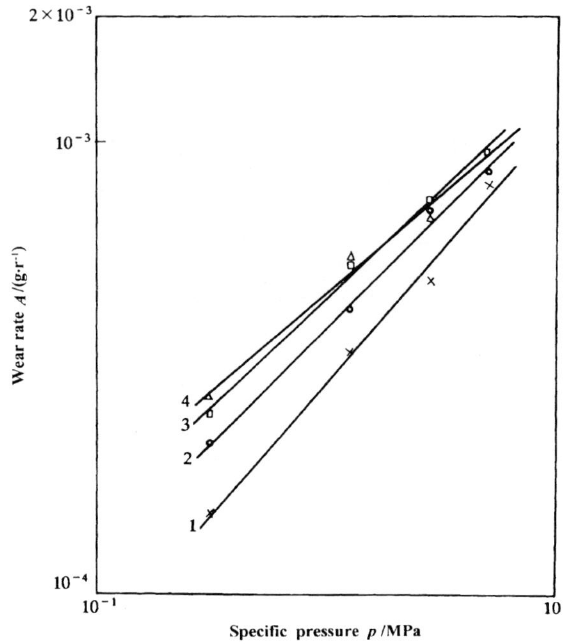

Fig. 4.41. Wear rates A plotted against specific pressure p (particle size 50 μm)

1 - υ=0.335m/s; 2 - υ=0.502m/s

3 - υ=0.670m/s; 4 - υ=0.837m/s

Along with the sliding speed increasing, the energy being ploughed the rubber surface by fixed particles is raised. Consequently, the wear rate is also increased exponentially, but by a smaller margin (Figure 4.42).

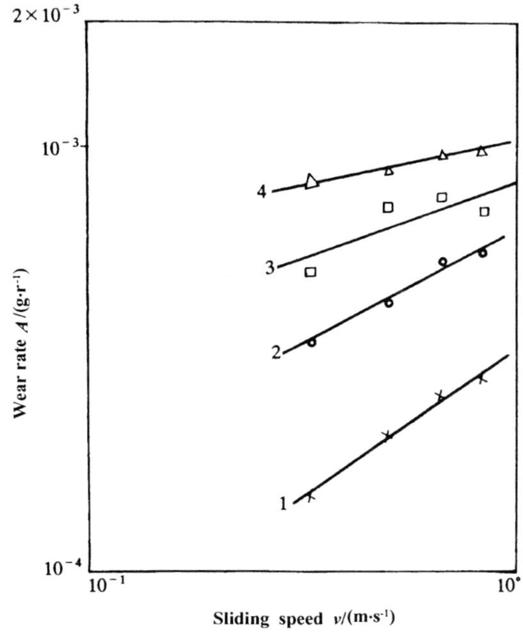

Fig. 4.42. Wear rates A plotted against sliding speed υ (particle size 50 μm)

1 – p=0.177 Mp_a ; 2 – p=0.354 Mp_a ;

3 – p=0.531 Mp_a ; 4 – p=0.708 Mp_a

The combined influence of specific pressure and sliding speed on the wear rate can be expressed as the following empirical equation:

$$A = kp^{\alpha}\upsilon^{\beta} \tag{4.21}$$

Where the coefficient k as well as the exponents α and β are dependent on the kind of rubber and the particle size d, for NBR material, these values are listed in Table 4.16.

Table 4.16. The coefficient k as well as the exponents α and β of NBR at room temperature

d, μm	k	α	β
50	1.742×10-3	1.027	0.459
45	0.542×10-3	0.576	0.370

A rise in particle size can increase the wear rates obviously (Figure 4.43). An empirical formula is given:

$$A = cd^t \tag{4.22}$$

Where the coefficient c and exponent t are dependent on the kind of rubber as well as the specific pressure and sliding velocity, for NBR material, their values are shown in Table 4-17.

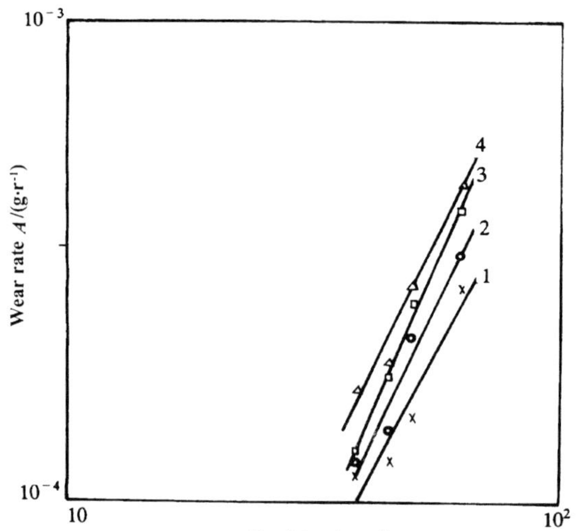

Fig. 4.43. Wear rates A plotted against particle size d (p=0.177MPa)
1— υ =0.335m/s; 2— υ =0.502m/s
3— υ =0.670m/s; 4— υ =0.837m/s

Table 4.17. The coefficient c and exponent t of NBR at room temperature (p=0.177MPa)

υ, m/s	c	t
0.335	9.87×10^{-8}	1.898
0.502	4.08×10^{-8}	2.175
0.670	1.89×10^{-8}	2.418
0.837	6.79×10^{-8}	2.124

Chapter 5

THEORY OF RUBBER ABRASION

In the past few decades, a number of scientists have studied the mechanism of rubber abrasion from various sides and obtained many valuable results. However, few general considerations can explicate quantitatively the experimental observation and discoveries in theory. In the strict sense, no substantial progress was made in this direction. To this end, the present author has made a serious thinking and active study to further facilitate the progress of the theory of rubber abrasion all along.

This chapter mainly introduces our theoretical achievements including some recent advances, which were obtained by applying some new theories and methods, such as fractal theory, computerized simulation technology and computer-performed analysis of image, in the study of the mechanisms and theory of rubber abrasion. Moreover, it will focus attention on the line-contact abrasion of rubber because its abrasion patterns closely resemble those produced on typical multi-asperity surfaces, as mentioned previous by Southern and Thomas [54].

5.1 Fatigue-Fracture Theory

5.1.1 Introduction

In 1974, Chomp, Southern and Thomas [66] proposed a simple theory relating the line-contact abrasion of rubber to the strength properties of rubber, in which the predicted relation between crack growth and abrasion was verified.

Later on, on the basis of experimental study and theoretical analysis, Southern and Thomas [54] presented a physical model for the steady rubber abrasion by a line contact (Figure 5.1) and a modification to the simple theory which describes the correlation between the volumetric wear-rate and frictional force as well as the crack growth behavior of rubber. In this theory, it is assumed that the abrasion pattern on the surface of the ring-shaped rubber sample (wheel) as stated previously is uniform, whereas the frictional force is exerted completely on the tongue of abrasion pattern. According to the fracture mechanics approach, the abraded volume of rubber per revolution of the rubber wheel, ΔV (cm^3/r), will be given by the following equation:

$$\Delta V = B[\frac{F(1+\cos\theta)}{b}]^{\alpha} Sb\sin\theta \tag{5.1}$$

Or

$$R_L = B[\frac{F(1+\cos\theta)}{b}]^{\alpha} \sin\theta \tag{5.2}$$

Where, F —frictional force;

S —circumference of rubber wheel;

b —width of rubber wheel;

B —constant related to crack growth behavior of rubber;

α —exponent related to crack growth behavior of rubber. It is equal to about 2 for natural rubber and to 4 or more for noncrystallizing unfilled rubbers, such as SBR [104].

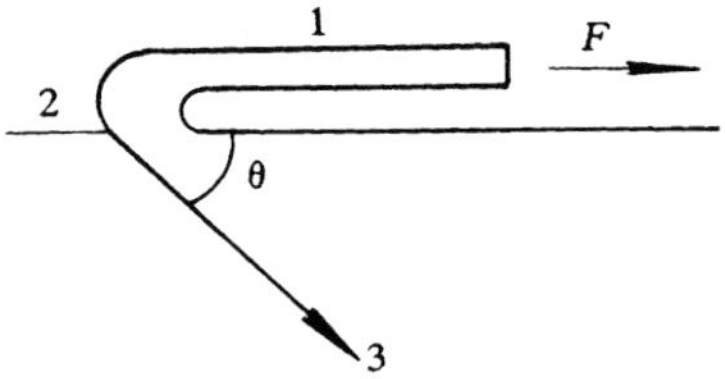

Figure 5.1. Physical model for crack growth
1-tongue; 2-rubber surface; 3-direction of growth of crack

Based on this simple theory, it is considered that the mechanism of rubber abrasion by a line contact can be ascribed to fatigue failure caused by the crack growth behavior of rubbers. The theory is supported by experimental results of some noncrystallizing unfilled rubbers. However, a series of experiments carried out by Gent and Pulford [55], using similar testing machine and test method, demonstrated that the simple theory in question has many limitations. For example, the wear-rates of natural rubber were found to be somewhat higher than for other rubbers even though its rates of crack growth are exceptionally low, especially under high stresses. The rates of crack growth for carbon-black-filled rubbers are not much lower than for unfilled rubbers and yet the wear-rates were greatly reduced. In addition, even though the temperature increment is able to increase the rate of crack growth by mechanical fatigue ever so much, a relatively small effect of temperature on the linear wear-rates of rubber abrasion was observed (Table 4.4). For instance, the rates of wear were found to be surprisingly insensitive to temperature, changing by a factor of about 3 at most between 25 °C and 100 °C, whereas much greater changes in rates of crack growth by mechanical fatigue are generally observed, by orders of magnitude for unfilled SBR. Therefore, Gent and Pulford considered that the mechanisms of rubber abrasion are partial mechanical fatigue and partial direct fracture under a single application of stress [55].

Years later, Gent [105] put forward a hypothetical mechanism for rubber abrasion being supplementary to the simple theory in question. This hypothetical is mainly involved with a mechanism of formation of the surface pits with a certain size of the order of a few μm under the action of frictional forces, namely, a mechanism of crack initiation in the subsurface of rubber during frictional sliding. The essentials of this hypothetical can be outlined as follows:

(1)Spalling of rubber surface by friction takes place as a result of subsurface fractures of rubber.

(2)The fractures of the subsurface layer of rubber are resulted from the microscopic precursor voids within the rubber. These voids produce unbounded elastic expansion and then burst open as cracks, under the action of internal pressure or of a triaxial tension in the surrounding rubber.

(3)The most probable mechanism of generating a sufficiently large inflation pressure or triaxial tension seems to be thermal decomposition of rubber.

This hypothetical mechanism could be used to explain better some questions which are unable to be clarified by the simple theory above. However, it is only a conjecture and needs to be improved by further experimental studies.

Fukahori and Yamazaki [63] also considered the viewpoint that the formation of abrasion pattern is only resulted from the crack-growth process by a single mechanical action to be unable to clarify the mechanism of rubber abrasion completely. Therefore, they proposed that the abrasion pattern is generated by two kinds of periodic motions, namely stick-slip oscillation and microvibration, as mentioned in section 4.1.2.

Based on the experimental studies of Gent and Pulford, the present author further carried out a series of experiments using the same tester in their laboratory, and put forward a theory of rubber abrasion by a line contact involving both unsteady and steady wearing stages [58]. Although this theory is still in consideration of the effect of the crack growth via mechanical fatigue on the linear wear-rates, the rupture of tongue tip of abrasion pattern resulted from tensile stress is considered as the direct cause of rubber wear. Over the years, this theory has been further evolved a more perfected fatigue-fracture theory of line-contact abrasion of the rubber after a deep going research. It can clarify better the whole process of rubber abrasion including unsteady and steady stages. The essential details of this theory will be given in the following sections.

5.1.2 Elastomechanical Analysis on the Origin of Abrasion Pattern

The basic feature of pattern abrasion of rubber is a ridged abrasion pattern occurred on the worn surface of the rubber. Therefore, to deeply reveal the origin and development of this abrasion pattern is the very core of a good understanding of the mechanism of rubber abrasion.

The physical model of interaction during contacting between the rubber and the razor blade is shown in Figure 2.15 of Chapter 2. The rubber surface is exerted by the normal and tangential loads during the razor blade sliding on the rubber surface. A certain site on rubber surface where the maximum shear stress is larger than the limiting shear strength of the rubber will generate cracks firstly. As known from the section 2.4, the frictional force of rubber friction by a line contact increases with an increase in the contact angle and changes with the sliding time periodically. Hence, if the contact angle reaches maximum, the frictional force will reach maximum either. Moreover, the cracks will be produced on the rubber surface unless the shear stress resulted from the frictional force is less than the limiting shear strength. Therefore, the origin of cracks is studied bellow by means of the analysis of stress distribution on rubber surface when the contact angle is a maximum.

In the beginning of 1950's, Smith and Liu presented a calculation method of contact stress of elastic solid under the combined action of both normal and tangential loads [106]. Here this method is applied to the analysis of stress distribution on rubber contact surface. Assuming that both the normal and tangential loads in the interface are elliptic distribution, then it can be analyzed being regarded as semi-infinite elastic solid under the action of elastic distribution load (Figure 5.2). The stresses on the contact surface (z=0) can be given by

$$\sigma_y = \begin{cases} -\dfrac{2}{3}p_0(\dfrac{y^2}{a^2}-1)^{1/2} & y \geq a \\ -p_0[(1-\dfrac{y^2}{a^2})^{1/2}+\dfrac{2y}{3a}] & y \leq a \\ -\dfrac{2}{3}p_0[\dfrac{y}{a}+(\dfrac{y^2}{a^2}-1)^{1/2}] & y \leq -a \end{cases} \tag{5.3}$$

$$\sigma_z = \begin{cases} -p_0(1-\dfrac{y^2}{a^2})^{1/2} & |y| \leq a \\ 0 & y \geq a \quad and \quad y \leq -a \end{cases} \tag{5.4}$$

$$\tau_{yz} = \begin{cases} -q_0(1-\dfrac{y^2}{a^2})^{1/2} & |y| \le a \\ 0 & y \ge a \quad and \quad y \le -a \end{cases} \tag{5.5}$$

Where, p_0 and q_0 are the maximum normal load and maximum tangential load corresponding to the origin of coordinates (i.e., y=0, z=0) respectively, and it is presumed that frictional coefficient $f=\dfrac{1}{3}$, thus $q_0 = \dfrac{p_0}{3}$; $2a$ is the width of contact point.

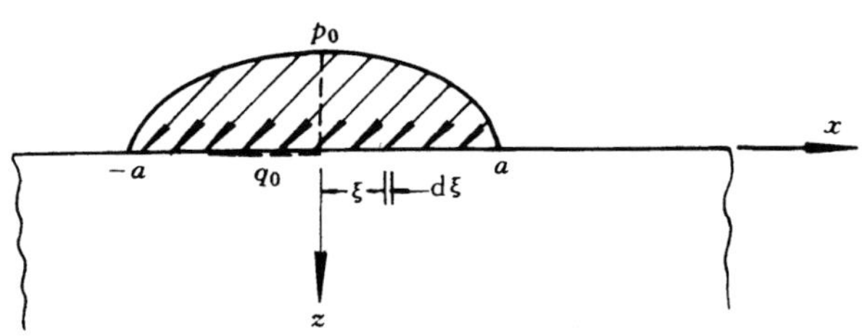

Fig. 5.2. Semi-infinite elastic solid exerted by elastic distribution load

The internal stress distribution of rubber induced by the distribution load is given by

$$\begin{cases} \sigma_y = \dfrac{-2q_0}{\pi z a}\displaystyle\int_{-a}^{a}(a^2-\xi^2)^{1/2}\dfrac{(y-\xi)^3 z}{[z^2+(y-\xi)^2]^2}d\xi \\ \qquad - \dfrac{2p_0}{\pi a z}\displaystyle\int_{-a}^{a}(a^2-\xi^2)^{1/2}\dfrac{z^2(y-\xi)^2}{[z^2+(y-\xi)^2]^2}d\xi \\ \sigma_y = \dfrac{2q_0}{\pi z a}\displaystyle\int_{-a}^{a}(a^2-\xi^2)^{1/2}\dfrac{(y-\xi)\ z^3}{[z^2+(y-\xi)^2]^2}d\xi \\ \qquad - \dfrac{2p_0}{\pi a z}\displaystyle\int_{-a}^{a}(a^2-\xi^2)^{1/2}\dfrac{z^4}{[z^2+(y-\xi)^2]^2}d\xi \\ \sigma_{yz} = \dfrac{-2q_0}{\pi z a}\displaystyle\int_{-a}^{a}(a^2-\xi^2)^{1/2}\dfrac{(y-\xi)^2 z^2}{[z^2+(y-\xi)^2]^2}d\xi \\ \qquad - \dfrac{2p_0}{\pi z a}\displaystyle\int_{-a}^{a}(a^2-\xi^2)^{1/2}\dfrac{z^3(y-\xi)}{[z^2+(y-\xi)^2]^2}d\xi \end{cases} \tag{5.6}$$

Integrating equation (5.6), the total contact stress of the semi-infinite elastic solid, which is produced by the elliptically distributive normal and tangential loads is given by:

$$
\begin{cases}
\sigma_y = \dfrac{q_0}{\pi}[(2y^2 - 2a^2 - 3z^2)\psi + 2\pi\dfrac{y}{a} + 2(a^2 - y^2 - z^2)\dfrac{y}{a}\overline{\psi}] \\
\quad - \dfrac{p_0}{\pi}z[\dfrac{a^2 + 2y^2 + 2z^2}{a}\overline{\psi} - \dfrac{2\pi}{a} - 3y\psi] \\
\sigma_y = \dfrac{q_0}{\pi}z^2\psi - \dfrac{p_0}{\pi}z(a\overline{\psi} - y\psi) \\
\sigma_z = \dfrac{q_0}{\pi}(a^2 + 2y^2 + 2z^2)\dfrac{z}{a}\overline{\psi} - 2\pi\dfrac{z}{a} - 3yz\psi] - \dfrac{p_0}{\pi}z^2\psi
\end{cases}
\tag{5.7}
$$

Where,

$$\psi = \frac{\pi}{k_1} \times \frac{1 - (\frac{k_2}{k_1})^{1/2}}{(\frac{k_2}{k_1})^{1/2}[2(\frac{k_2}{k_1})^{1/2} + \frac{k_2 + k_1 - 4a^2}{k_1}]^{1/2}} \tag{5.8}$$

$$\overline{\psi} = \frac{\pi}{k_1} \times \frac{1 + (\frac{k_2}{k_1})^{1/2}}{(\frac{k_2}{k_1})^{1/2}[2(\frac{k_2}{k_1})^{1/2} + \frac{k_2 + k_1 - 4a^2}{k_1}]^{1/2}} \tag{5.9}$$

$$k_1 = (a + y)^2 + z^2 \tag{5.10}$$

$$k_2 = (a - y)^2 + z^2 \tag{5.11}$$

Three main stresses σ_1, σ_2 and σ_3 can be obtained by means of Mohr circle:

$$\sigma_1 = \frac{\sigma_y + \sigma_z}{2} + [(\frac{\sigma_y - \sigma_z}{2})^2 + \tau_{yz}{}^2]^{1/2} \tag{5.12}$$

$$\sigma_2 = \frac{\sigma_y + \sigma_z}{2} - [(\frac{\sigma_y - \sigma_z}{2})^2 + \tau_{yz}{}^2]^{1/2} \tag{5.13}$$

$$\sigma_3 = \mu(\sigma_1 + \sigma_2) \tag{5.14}$$

Where, μ is Poisson's ratio.

Thus, the maximum shear stress will be given by

$$\tau_{\max} = \frac{\sigma_1 + \sigma_2}{2} \qquad \text{if } \sigma_1 > \sigma_2 > \sigma_3 \tag{5.15}$$

The maximum main stress σ_1 induced by the combined action of normal and tangential loads with elliptical distribution is shown in Figure 5.3, which is similar to the photoelastic stress distribution (Figure 4.3). The distribution of the equivalent lines of the maximum shear stresses induced by the combined action of both normal and tangential loads with elliptical distribution is shown in Figure 5.4. While a certain equivalent line reaches the limiting shear strength τ_0 of the rubber, then the crack will initiate and propagate along with the line. The angle included between the equivalent line and horizontal about ranges over $5^\circ \sim 25^\circ$, which is coincided with the experimental values [54,60]. As seen from Figure 5.3, under the action of identical load, the more the value of τ_{max}, the more the angle included between its equivalent line and horizontal line is. Hence, the crack angle reduces while the pressure (load) increases.

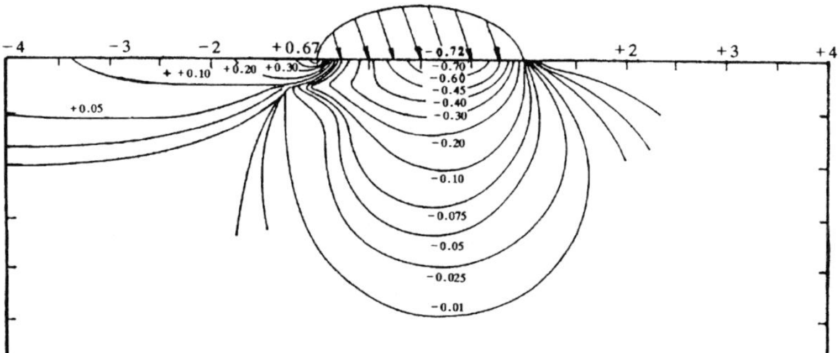

Fig.5.3. Maximum main-stress distribution under the combined action of both normal and tangential loads with elliptical distribution ($f=\frac{1}{3}$)

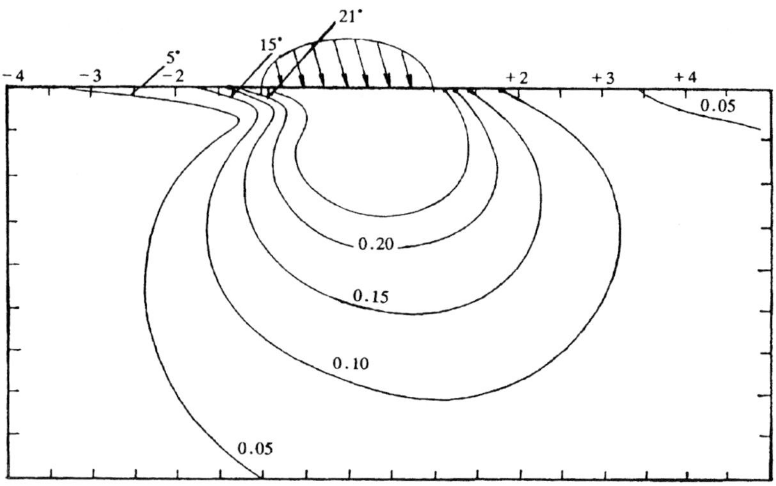

Fig.5.4. Maximum shear-stress distribution under the combined action of both normal and tangential loads with elliptical distribution ($f=\frac{1}{3}$)

As stated above, this analysis method can be not only applied to the stress analysis of rubber during the friction and wear processes, but also used to clarify the physical process of abrasion pattern formation of the rubber theoretically.

5.1.3 Physical Process of the Abrasion Pattern Formation

According to the elastomechanical analysis above and the experimental observations as stated in *chapter 4*, the physical process to form abrasion pattern might be considered as a periodically repeated processes as follows [58] (Figure5.5).

(a) During the process of razor blade sliding on the surface of rubber sample, the rubber surface is deformed and torn by the razor blade under the action of the normal load and the friction force. Consequently, a crack generates on the rubber surface at point A (Figure 5.5 (a), (b)), while the shear stress resulted from the frictional force is larger than the limiting shear strength of the rubber. After the crack initiates, a tongue forms on the surface. The razor blade comes into contact with the tongue and makes it bend backwards. Thus, a part of the rubber surface is protected from the scraping action of blade at the rear.

(b) Under the repeated action of tensile force produced by razor blade, on the one hand, the crack at point A is continuously propagated, and on the other hand, the fatigue rupture at the tip of tongue occurs at point B (Figure 5.5 (a), (b)). Then, the remained part of the tongue releases. Due to the tongue tip B being exerted tensile action, the crack at point A can be considered as crack type I and its crack angle is kept constant during propagating.

(c) Since the frictional force is changed with the sliding time periodically, it makes the tongue tip rupture gradually and peeling off completely from the rubber surface finally. Thus the ridged abrasion pattern with a rather uniform distribution is formed (Figure 5.5 (c)).

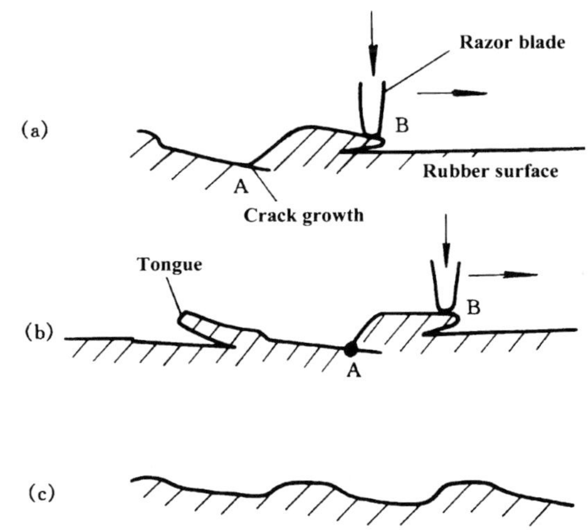

Fig.5.5. Processes of formation of abrasion pattern

Based on the above, a physical model of rubber abrasion involving both unsteady and steady wearing stages (Figure 5.6) and the corresponding physical model of the rupture process of tongue tip (Figure 5.7) were proposed [56, 58].

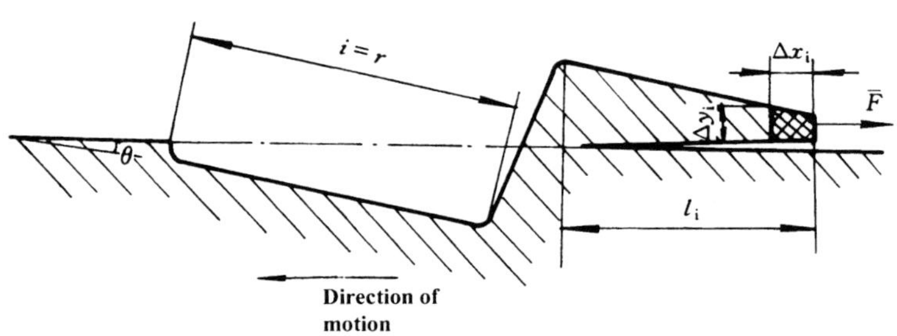

Fig.5.6. Physical model of rubber abrasion

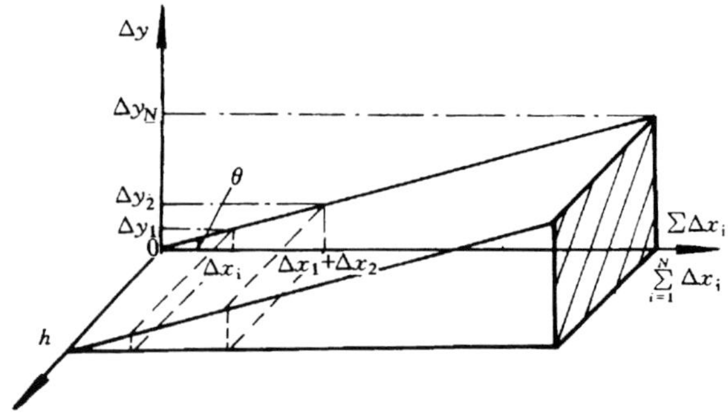

Fig.5.7. Physical model of rupture process of tongue tip

5.1.4 Mathematic Description of the Abrasion Pattern Formation [58]

In order to describe the process of the rupture of tongue tip precisely, a technical term, i.e. tensile rupture ratio ε_K, is introduced and defined as a ratio of the rupture length of the tongue tip Δx_i to the original length of the tongue l_i, as shown in Figure 5.6. Thus,

$$\varepsilon_K = \frac{\Delta x_i}{l_i} \qquad (i=1,2,\ldots) \tag{5.16}$$

As seen from Fig.5.7, we have

$$\Delta y_i = \sum \Delta x_i \cdot \tan\theta \tag{5.17}$$

For a given material and a constant frictional force, the angle θ can be considered approximately as a constant. Thus,

$$\Delta y_i \propto \sum \Delta x_i \tag{5.18}$$

Apparently, the rupture thickness of tongue tip has a maximum Δy_{max}. It is given by

$$\Delta y_{max} = \frac{\overline{F}}{\sigma_b} \tag{5.19}$$

Where, $\overline{F}$ is the frictional force per width; σ_b is tensile stress at break of tongue tip, namely tensile strength.

According to the above equations (5.17), (5.18) and (5.19), the tensile rupture ratio might be roughly considered to be directly proportional to the frictional force and inversely proportional to the tensile strength, namely

$$\varepsilon_K \propto \frac{\overline{F}}{\sigma_b} \tag{5.20}$$

Therefore, it is also supposed to be a constant for a given material under the condition of constant frictional force. Thus, $\Delta x_i \propto l_i$, obviously, the length l_i is mainly resulted from the crack growth and increased with the number of passes of the blade only. Moreover, as seen from Fig.5.6, r is the rate of crack growth, i.e., crack-growth length of rubber sample per revolution (pass), which is nearly a constant. In consideration of

$$l_i = ir - \sum_{i=1}^{i-1} \Delta x_i \tag{5.21}$$

and $\Delta x_0 = 0$, then the rupture process of the tongue tip can be described mathematically as follows.

After first pass (i=1),

$$\Delta x_1 = l_1 \varepsilon_K = r\varepsilon_K \tag{5.22}$$

After second pass (i=2),

$$\Delta x_2 = l_2 \varepsilon_K = (2r - \Delta x_1)\varepsilon_K = 2r\varepsilon_K - r\varepsilon_K{}^2 \tag{5.23}$$

After i-th pass (i=i), we have

$$\begin{aligned}\Delta x_i &= l_i \varepsilon_K = [ir - (\Delta x_1 + ... + \Delta x_{i-1})]\varepsilon_K \\ &= ir\varepsilon_K - [(i-1) + \frac{(i-1)(i-2)}{2!}]r\varepsilon_K{}^2 + \\ &[\frac{(i-1)(i-2)}{2!} + \frac{(i-1)(i-2)(i-3)}{3!}]r\varepsilon_K{}^3 - ...\end{aligned} \tag{5.24}$$

If the rubber abrasion process is just reached its stable stage after N revolutions, then the sum of rupture length of tongue tip is given by

$$\sum_{i=1}^{N} \Delta x_i = Nr\varepsilon_K f_0(N, \varepsilon_K) \tag{5.25}$$

$$\begin{aligned}f_0(N, \varepsilon_K) &= \frac{1}{2}(N+1) - [\frac{(N-1)}{2!} + \frac{(N-1)(N-2)}{3!}]\varepsilon_K + \\ &[\frac{(N-1)(N-2)}{3!} + \frac{(N-1)(N-2)(N-3)}{4!}]\varepsilon_K{}^2 - ...\end{aligned} \tag{5.26}$$

Obviously, N is the total number of revolutions or the sum of frictional distance corresponding with the critical point to transform the wearing state from unsteady to steady state; hence, it can be considered as a state criterion of rubber abrasion.

In general, the state criterion of rubber abrasion is a constant under otherwise identical conditions. It means the sum of frictional distance reaching a critical value, at this very moment, the wearing state of rubber abrasion is initially transformed from unsteady to steady state. Under the condition of Δx_i =r=1, an approximate representation as follows can be obtained by numerical calculation:

$$N = 1.2\varepsilon_K^{-0.9} \tag{5.27}$$

This state criterion might be applied to estimate the wear characteristics of various rubbers under similar running conditions.

5.1.5 Theoretical Wear Curve and Wear Equation of Rubber Abrasion [58]

5.1.5.1 Theoretical Wear Curve of Rubber Abrasion

The process of rubber abrasion might be generally divided into three stages and expressed correspondingly in terms of a theoretical wear curve (Figure 5.8) as follows.

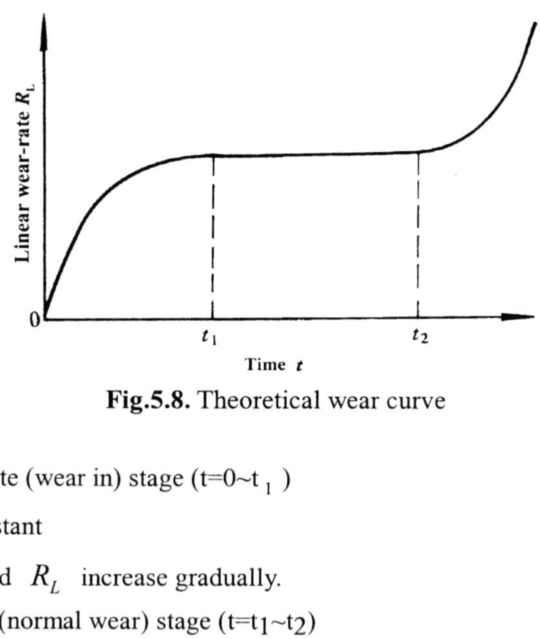

Fig.5.8. Theoretical wear curve

a. Unsteady state (wear in) stage (t=0~t_1)

Δx_i<r=constant

$\Delta x_i, \Delta y_i$ and R_L increase gradually.

b. Steady state (normal wear) stage (t=t_1~t_2)

Δx_i =r=constant;

$\Delta y_i = \Delta y_N \leq \Delta y_{max}$ =constant;

R_L =constant

c. Damage (accident wear) (t>t_2)

$\Delta x_i = r \neq$ constant ;

$\Delta x_i, \Delta y_i, r$ and R_L increase steeply.

5.1.5.2 Wear Equation of Rubber Abrasion (Linear Wear-Rate Equation)

Based on the physical process and the mathematic description of the abrasion pattern formation as well as related experimental results as stated above, the wear equation of rubber abrasion in unsteady state and steady state were obtained [58]:

In unsteady state

$$R_L = B(2\overline{F})^{\alpha} \sin\theta\eta(N, \varepsilon_K) \tag{5.28}$$

Steady state

$$R_L = B(2\overline{F})^{\alpha} \sin\theta \tag{5.29}$$

Where, $\overline{F}$ -frictional force per width;

B, α -constants being dependent on the crack-growth characteristics of materials;

$\eta(N, \varepsilon_K)$-characteristic function of wearing state of rubber abrasion.

$$\eta(N, \varepsilon_K) = \varepsilon_K f_0(N, \varepsilon_K) \tag{5.30}$$

Comparing equation (5.28) with equation (5.29), it is known that the value of characteristic function is always less than unity and approaches unity as a limit in the unsteady-state wear process, which is also coincident with the theoretical wear curve, namely, in unsteady-state stage, $0 < \eta(N, \varepsilon_K) < 1$. Apparently, while $\eta(N, \varepsilon_K) = 1$, it is corresponding with a critical point to transform the wearing state from unsteady to steady. Therefore, equation (5.28) can be considered as a general formula of rubber abrasion and equation (5.29) is only its an exceptional instance.

According to the results of numerical calculation of equation (5.30) on condition that r=1, two sets of theoretical curves of characteristic function (Figures 5.9 and 5.10) are obtained. Obviously, the value of the characteristic function is increased with an increase in the frictional distance i and tensile rupture ratio ε_K, which is in agreement with the experimental result [56]. As seen in Fig.5.10, the larger the tensile rupture ratio, the larger the value of the characteristic function is, and the earlier the steady state of wear is transformed (see Fig.5.9). This conclusion is also supported by the experimental observation that the larger the frictional force, the earlier the abrasion pattern formed.

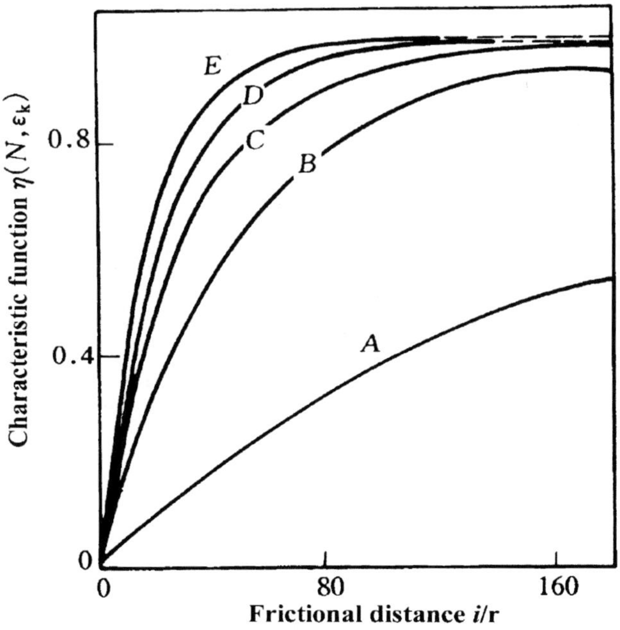

Fig.5.9. Characteristic function plotted against frictional distance.
A: ε_K =0.01; B: ε_K =0.02; C: ε_K =0.03; D: ε_K =0.04; E: ε_K =0.05

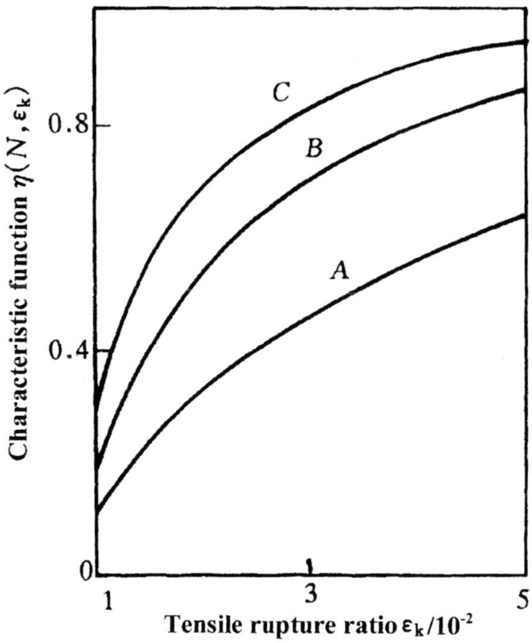

Fig.5.10. Characteristic function plotted against tensile rupture ratio (A. *i*=20rev; B. *i*=40rev; C. *i*=60rev)

As seen from equation (5.28), the linear wear-rates are mainly depended on the frictional work and changed exponentially. Moreover, the rates of wear also increase with an increase in frictional distance during unsteady state stage. These conclusions are in agreement with the empiric formula (4.8) and equation (4.9).

In addition, equation (5.28) not only comprehensively expresses the effects of various factors on the linear wear-rates, such as properties of the rubber and operating condition etc., but indicates the transformation condition of two wearing states as well.

Under the condition of steady-state wear, a simplified expression of the wear equation can be obtained:

$$R_L = k(2\overline{F})^{\alpha} / \sigma_b \qquad (5.31)$$

Where, k is a coefficient.

The above equation reveals further the relationship between the rates of wear and the properties of materials. Thus it can be seen that the reason why filling carbon black can reduce notably the wear-rate of amorphous elastomers at room temperature might be resulted from an increase in the tensile strength of the carbon-black-filled rubber.

5.2 Energy Theory

5.2.1 Energy Theory of Metal Wear

The energy theory of wear is firstly presented by G.Fleischer [107]. He proposed that friction is a process of energy distribution, which transforms the energy into the phenomena of heat, noise, electricity and wear of material and so on, moreover, the work performed by the frictional force is equal to the total of the energy losses above (Fig5.11).

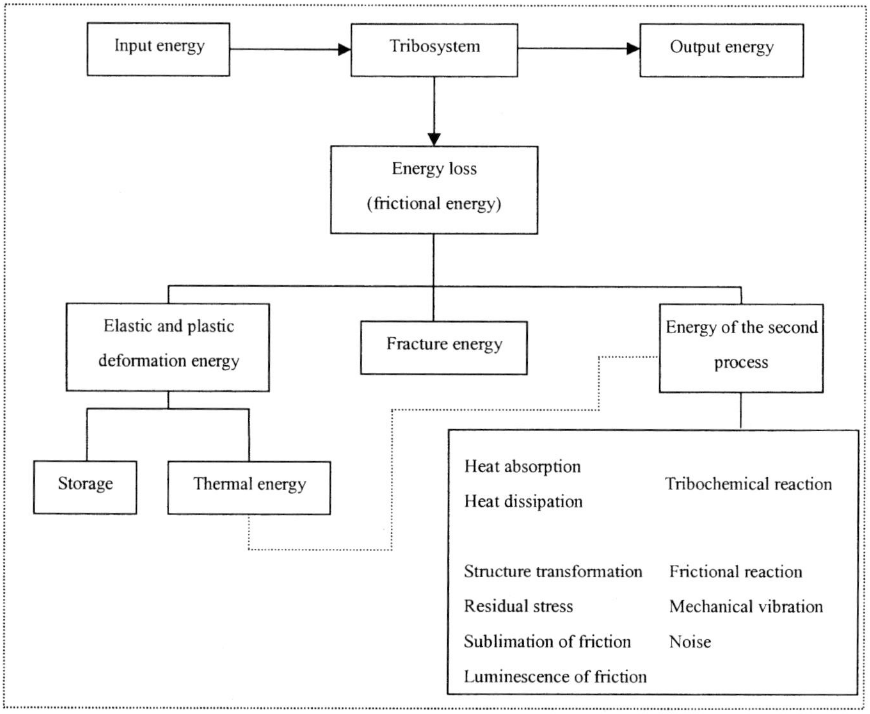

Fig 5.11. Transformation, distribution and interaction of the energy in tribological process

Fleischer has further pointed out that under the action of the frictional work in the friction process; every local volume of the frictional surface generates deformation. The frictional work is dissipated mostly in the form of heating and a fraction of it (about 9%-16%) is accumulated in each partial volume in the form of potential energy. Once the potential energy in a certain particle volume of the surface accumulates to a critical value enough to cause the surface to damage, the material of this local volume will be separated from the surface in the form of wear debris, namely the wear of material occurs. It is shown that friction is almost always induced the wear of materials.

In order to connect friction with wear in the view of energy, Fleischer [107] introduced a concept of hypothetic frictional energy density. It means the consumption of frictional work of unit volume of that material to be worn off exerted by the frictional force one time, namely:

$$e_R^* = \frac{W_f}{V} \tag{5.32}$$

Where, e_R^* - hypothetic frictional energy density;

W_f - frictional work;

V - wearing volume of material caused by the frictional work.

In addition, he also introduced a concept of basic energy density, which is referred to the consumption of frictional work of unit volume of that deformed part applied by the frictional force one time:

$$e_{R_e} = \frac{W_f}{D_d} \tag{5.33}$$

Where, e_{R_e} - basic energy density;

D_d - deformed volume.

5.2.2 Energy Theory of Rubber Abrasion

5.2.2.1 Wear Energy Density

Although the concept of frictional energy density presented by Fleischer relates the frictional work with the wearing volume during the friction and wear processes of materials, only partial frictional work is converted into the potential energy of material to cause wear of material, and the converting ratio is mainly relative to the practical physical process. So the two kinds of energy density proposed by Fleischer can not reveal the nature of wear of material. Hence, this energy theory is only enabled to explain qualitatively the wear phenomenon. In fact, the true reason for producing the material loss and wear debris lies in that a considerable potential energy is accumulated on the surface of material. Therefore a concept of wear energy density e^* is presented. The physical meaning of this new concept is the accumulated internal energy being required to generate the wear of material of unit volume:

$$e^* = U_{in} / V \qquad .34)$$

Where, V - wearing volume of material;

U_{in} - internal energy accumulated in the wearing volume.

This wear energy density is a constant related to the material, which can be used to describe quantitatively the wear resistance of material.

5.2.2.2 Wear Equation [108]

Based on the results of experimental studies of the present author, as stated in Section 4.1.2, the physical process of rubber abrasion might be mainly considered as two alternatively proceeded processes i.e., crack growth of surface layer (tongue formed) and rupture of tongue tip of ridges (tongue ruptured) (see Fig4.11). The latter is the direct cause of material losses although the effect of crack growth via mechanical fatigue on abrasion is also taken into account. Hence, a correspondent physical model similar to Fig.5.6 is shown in Fig.5.12, which furnishes the basis of the following analysis.

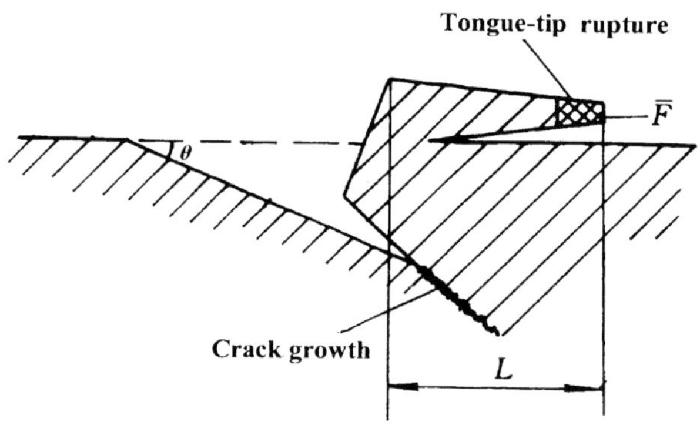

Fig.5.12. Physical model of rubber abrasion by a line contact

As mentioned above, only a fraction of the frictional work could be converted into the rupture energy during wearing process. The accumulated internal energy can be expressed as the frictional work which is done in the two processes of formation and rupture of tongue as follows:

$$U = KPfL \tag{5.35}$$

Where, U - accumulated energy;

K - constant related to the property of material and the accumulation of internal energy;

P – normal load;

f – coefficient of friction;

L - length of the crack-growth (tongue).

According to the theory of fracture mechanics and from the viewpoint of energy, the accumulated energy could be considered to consist of crack growth energy and rupture energy of the tongue tip. Thus,

$$U = U_A + U_B \tag{5.36}$$

Where, U_A - crack - growth energy;

U_B - ruptured energy of the tongue tip.

However, the crack growth per stress cycle, namely the rate of crack growth ***r*** is given by [54]:

$$r = BT^{\beta} \tag{5.37}$$

Where, B - constant;

T - rupture energy;

β - constant related to material, for NR, $\beta = 2$;

and

$$T \approx 2\overline{F} \tag{5.38}$$

$$\overline{F} = F / a \tag{5.39}$$

Where, $\overline{F}$ - frictional force per width;

a - contacting width between the scraper and the surface of rubber sample.

Thus,

$$r \approx B(2\overline{F})^{\beta} \tag{5.40}$$

As seen in *Chapter 4*, the crack growth per stress cycle is a constant for the same material under the action of identical frictional work. The energy conversion, namely converting elastic energy kept the propagating of crack to surface energy, is proportional to the length of crack propagation but nothing to the shape of sample and the loading mode. Therefore, the energy for crack growth $\boldsymbol{U_A}$ is constant while the length of crack-growth is unchangeable. From equations (5.35) and (5.36), we have

$$U_B = U - U_A = KPfL - U_A \tag{5.41}$$

From the equation above, it can be seen that in the early wearing stage, as the length of tongue is shorter and the corresponding $\boldsymbol{U_B}$ is smaller, the rate of rupture will be lower than the rate of crack-growth. In the process of wearing, the tongue is torn longer and longer, and then the $\boldsymbol{W_f}$ and the $\boldsymbol{U_B}$ are increased. Consequently, the rupture rate of tongue rises more rapidly, and when it is equal to the rate of crack-growth, the steady stage of wear is reached.

Now the concept of wear energy density as mentioned above is introduced to further analyze the wearing process. According to equation (5.34), we have

$$e^* = U_B / V_B \tag{5.42}$$

Where V_B is the ruptured volume per stress cycle, namely

$$V_B = S\Delta L_n \tag{5.43}$$

Where, S - cross section of the ruptured tongue tip;

L_n - rupture length of the tongue tip at the nth stress cycle.

From equations (5.35) and (5.36), we have

$$U_A = U - U_B = KPfL - U_B \tag{5.44}$$

Inserting equations (5.42) and (5.43) into the above equation, we obtain

$$U_A = KFL_n - e^* S\Delta L_n \tag{5.45}$$

In light of that the wear process might be considered as a continuous process and the crack-growth energy U_A is constant under the identical operating condition, then

$$U_A = KPfL_n - e^* S\Delta L_n = KPfL_{n+1} - e^* S\Delta L_{n+1} = \cdots \tag{5.46}$$

Where L_n and L_{n+1} are the length of crack growth at the ***n***th and $(n+1)$th stress cycle respectively, ΔL_n and ΔL_{n+1} are the rupture length of tongue tip at the nth and $(n+1)$th stress cycle, respectively.

Thus

$$KPf(L_{n+1} - L_n) = e^* S(\Delta L_{n+1} - \Delta L_n) \tag{5.47}$$

Based on the experimental observation, it was assumed that a rupture fragment is produced for each stress cycle [58].

Then

$$L_{n+1} - L_n = r - \Delta L_n \tag{5.48}$$

Inserting equation (5.48) into equation (5.47), we have

$$\Delta L_{n+1} - \Delta L_n = \frac{KPf(r - \Delta L_n)}{e^* S} \tag{5.49}$$

The above equation can be rewritten as

$$\frac{d\Delta L_n}{dn} = \Delta L_{n+1} - \Delta L_n = \frac{KPf(r - \Delta L_n)}{e^* S} \tag{5.50}$$

Introducing the beginning conditions (n=0, L_n=0), the rupture length of the tongue tip per stress cycle can be obtained by integration:

$$\Delta L_n = r - r\exp(-\frac{kPfn}{e^* S}) \tag{5.51}$$

Therefore, the linear wear - rate of abrasion by a line contact in unsteady state can be given:

$$R_u = \Delta LS \frac{D}{S} \frac{1}{Da} \tag{5.52}$$

Where, D - distance of friction per revolution;

S - average spacing of ridge;

a - contact width of scraper;

$$S = ha \tag{5.53}$$

Where, h is the thickness of tongue.

$$h \approx \overline{S}\sin\theta \tag{5.54}$$

Where θ is the crack - growth angle.

Inserting equations (5.53) and (5.54) into equation (5.52), gives

$$R_u = [r - r\exp(-\frac{KPfn}{e^* S})]\sin\theta \tag{5.55}$$

then

$$R_u = r\sin\theta - r\exp(-\frac{KPfn}{e^* S})\sin\theta \qquad 5.56)$$

The above equation could be rewritten as:

$$R_u = r\sin\theta[1 - \exp(-\frac{KPfn}{e^* S})] \tag{5.57}$$

However, the theoretical wear-rate equation of abrasion by a line contact in steady state was derived previously [54,58], that is:

$$R_s = r\sin\theta \tag{5.58}$$

and the general expression is:

$$R_s = B(2\overline{F})^{\alpha}\sin\theta \tag{5.59}$$

Inserting equation (5.58) into equations (5.56) and (5.57) respectively, the final expressions are given by

$$R_s - R_u = r\exp(-\frac{KPfn}{e^* S}) \tag{5.60}$$

and

$$\frac{R_u}{R_s} = 1 - \exp(-\frac{KPfn}{e^* S}) \tag{5.61}$$

As seen, from equation (5.60), the logarithms of the difference between the wear rate in unsteady state and that in steady state is inversely proportional to the number of stress cycles, the normal load and the frictional coefficient, but is proportional to the wear energy density and the cross – section of the tongue tip.

From equations (5.35), (5.36), (5.42) and (5.43), without considering the energy for the crack growth of tongue root, namely U_A=0, then

$$\Delta L_n / L = Se^* / FK \tag{5.62}$$

Where, L_n/L is the very tensile rupture ratio of rubber abrasion ε_K defined in Section 5.1.4. As seen from the above equation, in case of unchanging working condition, $\varepsilon_K \propto e^*$, which proves that the result obtained by energy theory is consistent with that obtained by fatigue – fracture theory under the condition of without consideration on the energy required for the crack growth of tongue root.

5.2.2.3 Experimental Verification

In order to check the correctness of the equations above, experiments of abrasion by a line contact for NB, SBR and NBR materials were carried out with the testing method as described in Section 2.4.1.

(1) NR

The experimental correlation of the wear rate and the number of revolutions is shown in Figure 5.13. The wear rate increases with the number of revolution and keeps constant once the number of revolutions rises to 1800r.

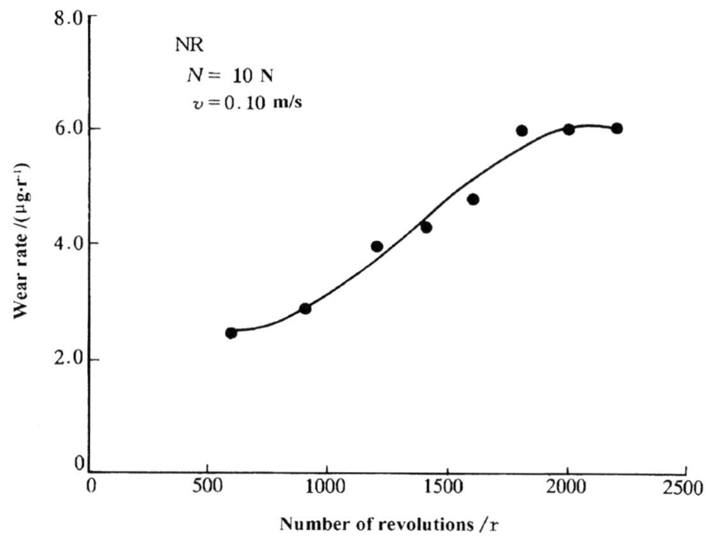

Fig.5.13. Linear rates of wear of NR material plotted against the number of revolutions in unsteady state (N=10N, *v*=0.10m/s)

The difference between the wear rates in steady state and that in unsteady state is taken as logarithm and shown in Figure 5.14. It is clear that the difference of experimentally determined wear rates between the two kinds of abrasion state is linear with the number of revolutions, which is in accord with equation (5.60).

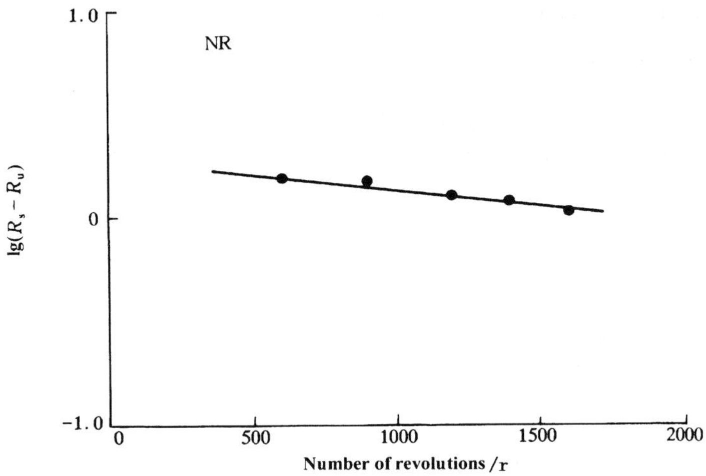

Fig 5.14. The logarithm of difference ($R_S - R_u$) of NR material plotted against the number of revolutions

(2) SBR

The variation of the wear rate with the number of revolutions is illustrated in Figure 5.15. The wear rate is increased with the number of revolutions. While the number of revolutions is up to 400r, the wear rate is unchangeable.

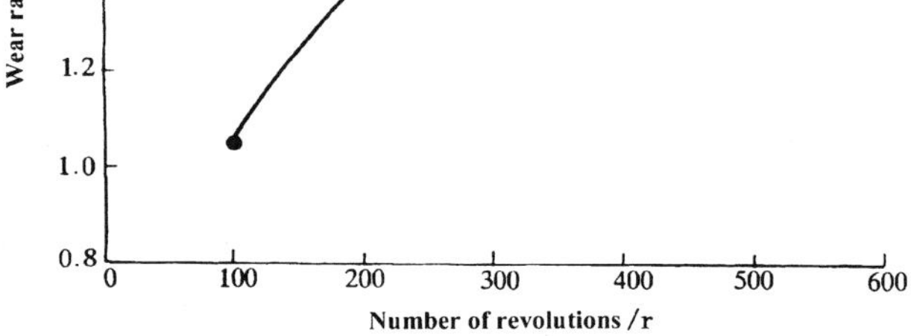

Fig.5.15. Linear wear-rates of SBR material plotted against the number of revolutions (N=8N, v=0.10m/s)

The difference between the wear rates in steady state and that in unsteady state is taken as logarithm and shown in Figure 5.16. As seen, the difference of experimentally determined wear rates between the two kinds of abrasion state is also directly proportional to the number of revolutions.

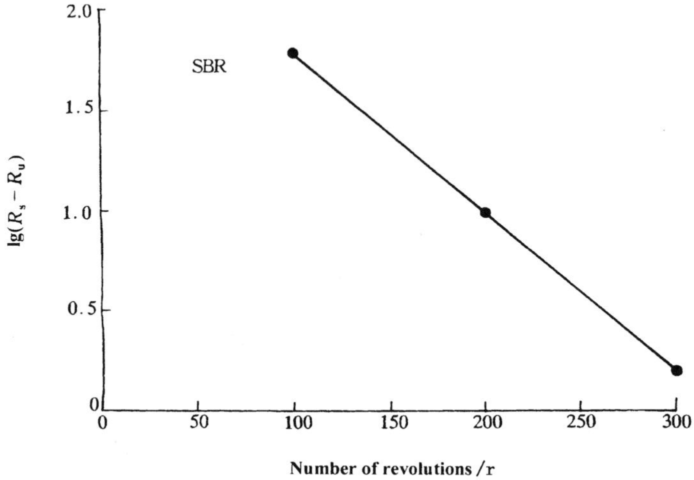

Fig. 5.16. The logarithm of differences ($R_S - R_u$) of SBR material plotted against the number of revolutions

(3) NBR

The experimental relation of the wear rate and the number of revolutions is shown in Figure 5.17. The wear rate is increased with the number of revolutions. After the number of revolutions goes to 1800r, the wear rate will be unchanged.

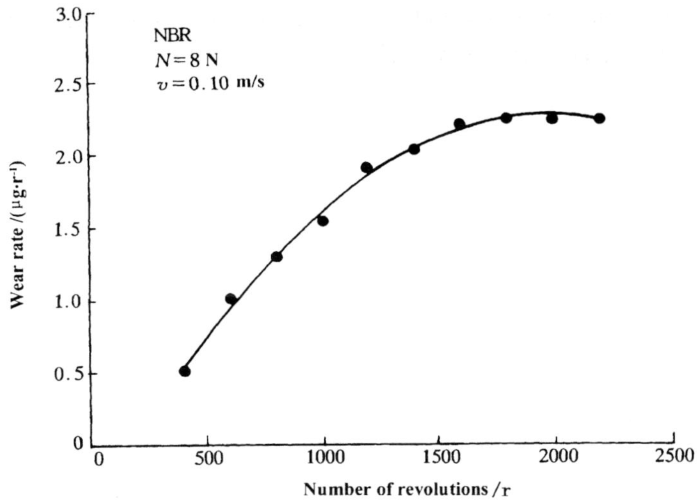

Fig.5.17. Linear wear-rate of NBR material plotted against the number of revolutions (N=8N,V=0.10m/s)

The difference between the wear rates in steady state and that in unsteady state is shown logarithmically in Figure 5.18. As seen, the difference of experimentally determined wear rates between the two kinds of abrasion state is almost linear with the number of revolutions.

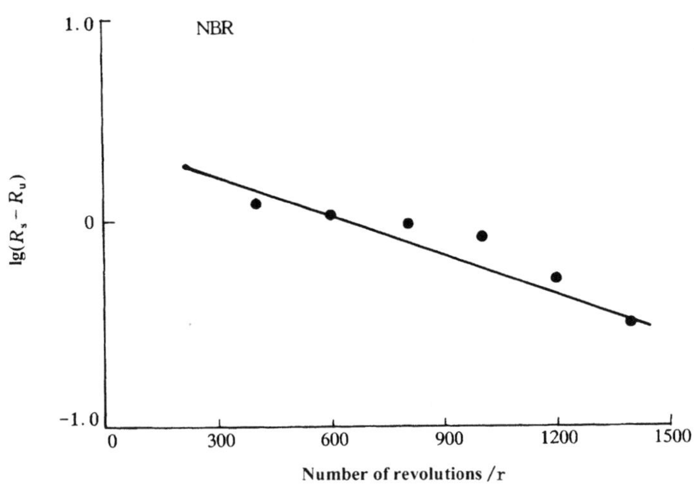

Fig.5.18. The logarithm of differences (R_s-R_u) of NBR material plotted against the number of revolutions.

The evidence from the experiments in question proves the correctness of the energy theory of rubber abrasion qualitatively. Moreover, from the relations of the wear rates of NR, SBR and NBR materials with the corresponding number of revolutions, it is also found that the wear rate in unsteady state is increased with the number of revolutions.

5.3 Fractal Analysis and Computer Simulation of Wear Surfaces

"Fractal" is a technical term fathered by B.B.Mandelbrot in 1975. Its primary meaning is an irregular, fractional and fragmented object. It refers to a kind of much irregular and disordered complex system, but a part of which is similar to the whole in some way. This similarity can be described quantitatively with a parameter, i.e., dimension of fractal. The formation process of this system is random. So, fractal theory is a new subject for studying the unity of order and disorder, definition and randomness in the nonlinear system. Though it only has a very short history of growth and is still in developing phase now, it has presented a new thought and method for modern science and technology. Therefore, of recent years this theory was introduced to many fields more and more extensively, thus pushing these areas forward.

Late in the 1980's, fractal theory began to be applied to the study in rubber abrasion. Based on the brief introduction of related work being carried out at abroad, this section is mainly introduce the prime results that we had obtained by applying the fractal theory to the analyses of abrasion pattern and wear debris of rubber abrasion of late years.

Stupak and co-workers [109-111] were applied the fractal method to analyze the worn surfaces and wear debris of rubber for the first time. They conducted the experiment of abrasion by using a line-contact abrasion abrader as shown in Fig4.9. Three rubber compounds including NR, PBD (polybutadiene) and SBR materials were tested under different frictional work ranging from 650 to 2160 J/m^2.

The essence of fractal analysis is that the length (or area) of an irregular line (or surface) depends on the size of the measuring device. A graphical representation of log length L of a mathematical fractal curve against log measuring unit size R yields a straight line described by the following relation:

$$L \propto R^{1-D} \tag{5.63}$$

Where D is the fractal dimension.

5.3.1 Fractal Analysis of Wear Surfaces

Stupak and co-workers [109] obtained the surface profiles roughness of worn surface by profilometer and mounted the profilometer traces on a paper board with 1mm thickness, then rolled discs of various diameters along the length of the trace edge to get the relation between $\boldsymbol{L}$ and $\boldsymbol{R}$. From Fig5.19, it can be seen that $\boldsymbol{L}$ is linear with $\boldsymbol{R}$ in a certain scale (region B), which indicates the worn surface of rubber being characterized by fractal only in this scale (region B), and its fractal dimension can be determined from equation (5.63).

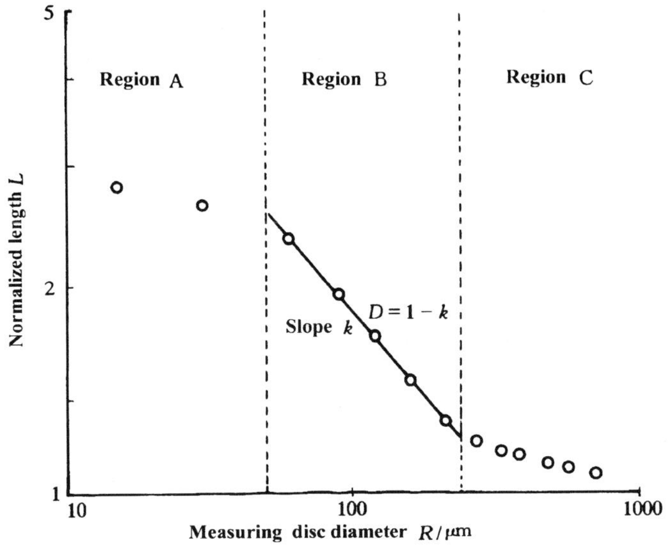

Fig 5.19. Typical fractal plot of profilometry data [109]

Comparing the fractal curves of same material (e.g., NR material) under different input work (Figure 5.20), it is shown that when D was constant with frictional work or the wear mechanism is unchangeable, the fractal curves could be superimposed over their entire lengths (regions A, B and C) by shifting the curves only along the measuring unit axis (Fig 5.21). Similar results were obtained for PBD and SBR materials in Reference [109]. Stupak et al [110~111] further investigated the fractal dimension in relationship with the wear mechanism, load and velocity, and found that the fractal dimension has nothing to do with the frictional force and velocity for the same wear mechanism.

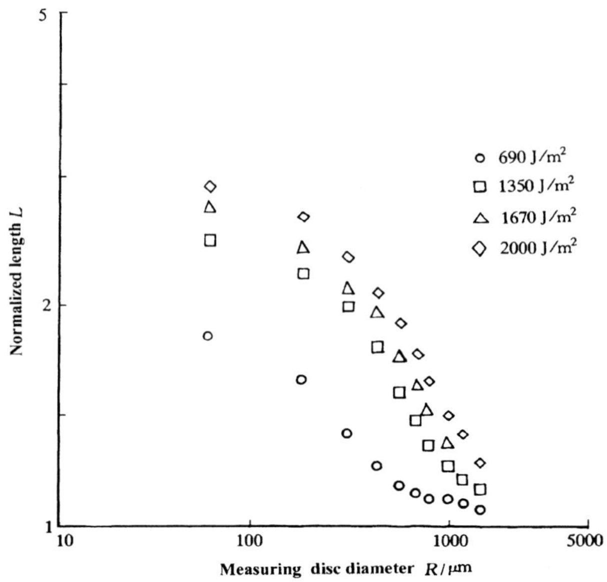

Fig 5.20. Fractal plot of profilometry data for worn surfaces of NR [109]

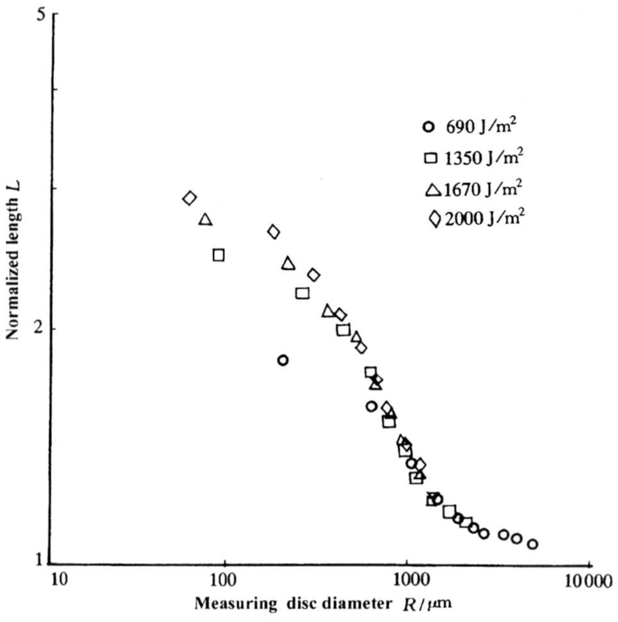

Fig 5.21. Shifted fractal plot for worn surfaces of NR [109]

5.3.2 Fractal Analysis of Wear Debris

Through measuring the perimeter P and the projected area A of each wear particle at various frictional work, Stupak and Donovan [109] had found that the perimeter P is linear with the projected area A of the debris on the log-log coordinate under the condition of constant frictional work (Fig.5.22), which shows that the debris is also fractal, and its fractal dimension can be defined by the following equation:

$$P \propto A^{0.5D_d} \tag{5.64}$$

Where, A - projected area of debris;

D_d - fractal dimension of debris.

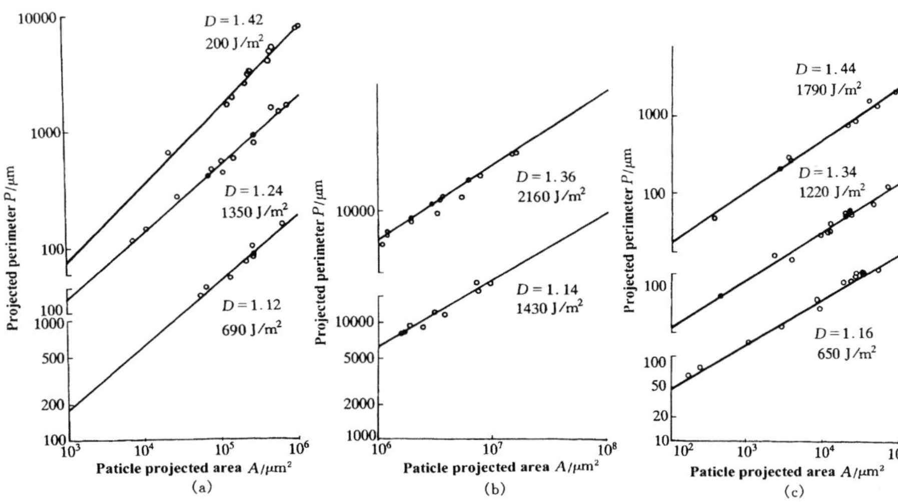

Fig 5.22. Fractal plots of rubber debris [109]
(a) NR (b) SBR (c) PBD

They also found that the fractal dimension $\boldsymbol{D_d}$ of the wear debris is proportional to the normal load.

From above, Stupak and co-worker had put forward the method for calculating the fractal dimension of the worn surface of rubber abrasion and revealed preliminarily the correlation between the fractal dimension and the normal load, but they did not find out the intrinsic relation between the fractal dimension of worn surface and the characteristics of wear process. Thereby they cannot clarify the physical meaning of the fractal dimension of worn surface in the view of tribology.

5.3.3 Fractal Calculation of the Crack Angle of Abrasion Pattern

As seen in Section 4.1.2 of *Chapter 4*, the values of crack angle have great influence on the generation and development of abrasion pattern. Therefore, to determine the value of crack angle is of great importance to reveal the mechanism of rubber abrasion.

Southern and Thomas [54], and Uchiyama and Ishino [60] applied same method to calculate the value of crack angle. They assumed that the abrasion pattern on the rubber surface moved forward evenly in the process of formation (Figure 5.23) and considered that the value of crack angle is related to the moving speed of the abrasion pattern and the linear wear rate of rubber. Thus the crack angle can be determined by measuring the moving speed of the abrasion pattern and the linear wear rate of rubber. Obviously, this method is much overelaborated.

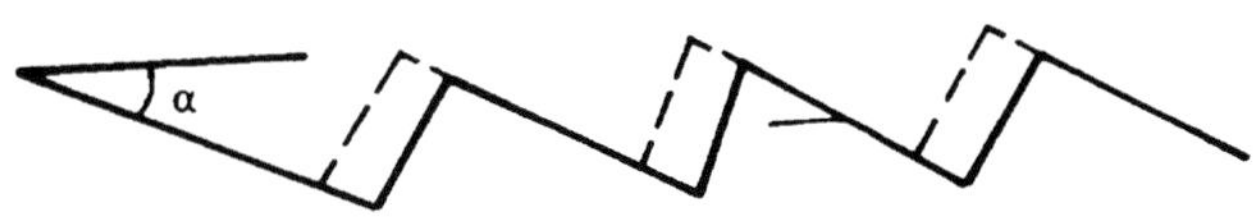

Fig 5.23. Physical model of abrasion pattern movement

5.3.3.1 Relation of Fractal Dimension and Value of Crack Angle

In the early 1980's, the present author [57] observed for the first time that the abrasive pattern on the surface was grown from small to large with similar morphology in the developing process of rubber abrasion from unsteady state to steady state, namely, the gradational overlap with self-similarity was formed (see Section 4.1.2). In the light of the observation and analysis of the primary and secondary abrasion patterns on the worn surface of rubber (Figure 5.24), a generating element model of rubber abrasion is presented, as shown in Fig 5.25.

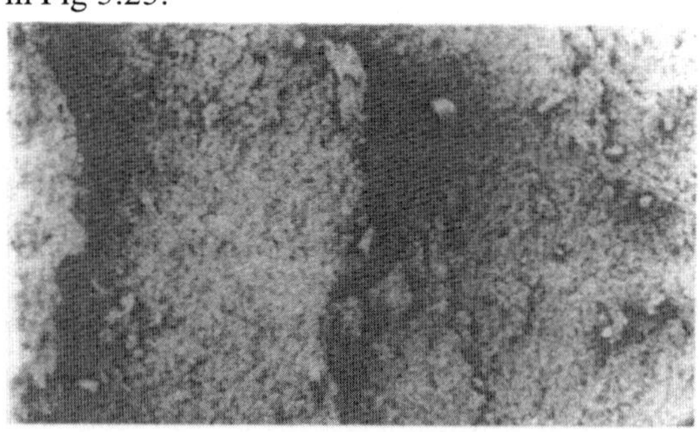

Fig. 5.24. Primary and secondary abrasion patterns on the worn surface of rubber

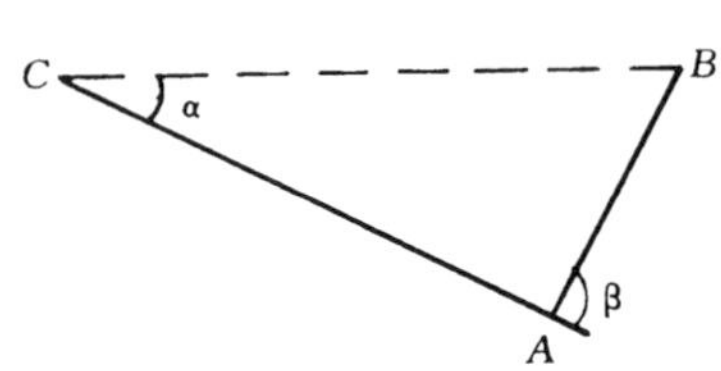

Fig.5.25. A generating element model of rubber abrasion

Taking AB=1,CA=k, then the number of measuring Ni is given by

$$Ni = k + 1 \tag{5.65}$$

The similar ratio r is expressed as [112]

$$1/r = (k^2 + 1 + 2k\cos\beta)^{1/2} \tag{5.66}$$

Then the dimension D is [112]

$$D = \lg Ni / \lg(1/r) = \lg(k+1)/\lg(k^2 + 1 + 2k\cos\beta)^{1/2} \tag{5.67}$$

In the case of β=90°, k=cot , we have

$$D = 2\lg(k+1)/\lg(k^2 + 1) \tag{5.68}$$

namely,

$$D = 2\lg(\cot\alpha + 1)/\lg(\cot\alpha^2 + 1) \tag{5.69}$$

As seen from the above equation, a correspondence relation between the fractal dimension D and the crack angle α is existed (Fig. 5.26). Consequently, the crack angle can be determined by calculating the fractal dimension of the worn surface.

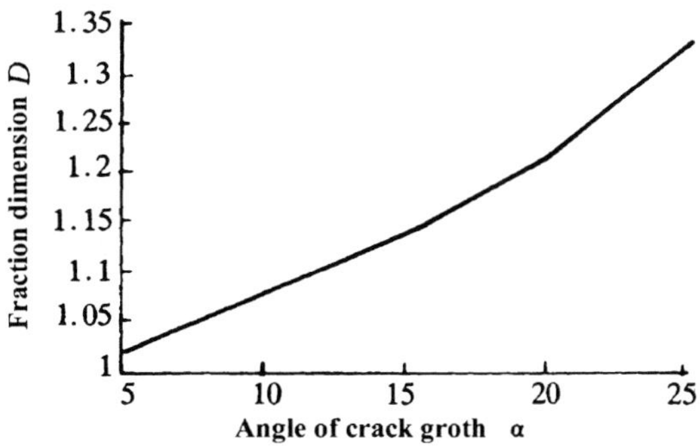

Fig. 5.26. Fractal dimension v_s crack angle

5.3.3.2 Method of Calculating the Fractal Dimension of the Worn Surface

The variation method on the basis of covering principle [113] is used to calculate the fractal dimension. The advantage of this method is easy to carry out with computer. The calculating process is described as follows. An initial measuring unit size R_i is selected as the horizontal distance which includes a certain number of digitized data points. From the beginning of trace, a rectangle is created with width R_i and height Hij, which is the difference between the maximum and minimum of digitized points within the interval R_i. The area of this rectangle is expressed as Sij. With the endpoint of the rectangle as the beginning of the next one and repeating the above calculation until the endpoint of the trace is reached (Fig.5.27). If the number of rectangles obtained is n_i, we have

$$Si = \sum_{j}^{n_i} Sij \tag{5.70}$$

And the number of measurements can be given by

$$N_i = S_i / R_i^2 \tag{5.71}$$

Based on the fractal principle, we have

$$N \propto (1/R)^D \tag{5.72}$$

$$\lg N \propto (-D)\lg R \tag{5.73}$$

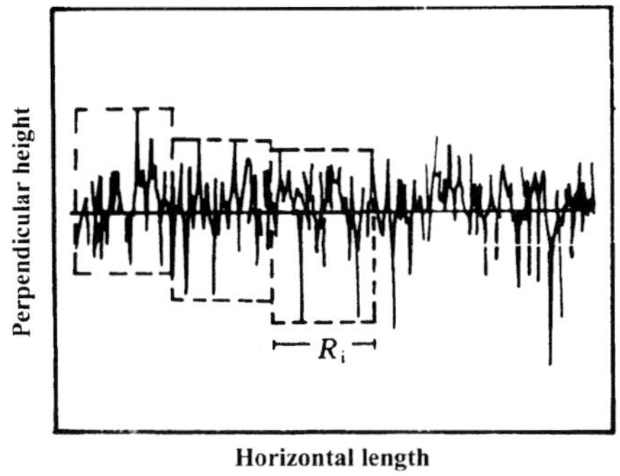

Fig.5.27. Scheme of variation method

If Ni - Ri relation of a worn surface of rubber shown on the lg-lg coordinate is a straight line, namely $\lg N_i$ is linear with $\lg R_i$, this worn surface is fractal, and its fractal dimension D_f is equal to the negative value of the slope of the line.

Therefore, the fractal dimensions can be determined directly by the slope of the regression line on the lg-lg coordinate. Because the surface of material is statistically self-similar in general and the judgment on fractal is formed by the line which is accumulated by the data points on plane, it is necessary to find a way to judge if the data points gather in line, and then to estimate the fractal dimension.

The Ri-Ni relation is obtained by the variation method and then the coordinates of several points are gained on the lg-lg coordinate respectively: (x_1,y_1), (x_2,y_2),...,(x_n,y_n).

Assuming

$$l_{xx} = \sum (x_i - \bar{x})^2 \tag{5.74}$$

$$l_{yy} = \sum (y_i - \bar{y})^2 \tag{5.75}$$

$$l_{xy} = \sum (x_i - \bar{x})(y_i - \bar{y}) \tag{5.76}$$

Then the equation of regression line is given by

$$y = \alpha + \beta \tag{5.77}$$

Where,

$$\beta = l_{xy} / l_{xx} \tag{5.78}$$

$$\alpha = \bar{y} - \beta x \tag{5.79}$$

Later on, reducing gradually the number of the data points and then continually regressing until the values of the slope calculated before and after are almost equal, thus it can be thought that y is linear with x in this region which is considered as nonscale, and the fractal dimension is equal to the negative value of the slop -*β*.

In order to prove the correctness of this method and its corresponding procedure, the equation of Weierstrass-Mandelbrot (W-M equation)[113] is chosen to simulate the surface of material, and then the corresponding fractal dimension is calculated by the variation method.

Strictly speaking, the surface of material is not continuous as view from the atomic scale, but it could be considered as continuous and the tangent corresponding to each point is not existed in the view of engineering. Therefore, every point on the surface of material might be regarded as continuous, but cannot be differentiated, so did the W-M equation, as a result, this equation can be used to describe the engineering surface of material:

$$z(x) = \sum_{n=1}^{\infty} b^{-nH} (1 - \cos b^n x) \tag{5.80}$$

Where, b>1, 0<H<1, D=2−H.

It is given as H=0.6, b=2.1, x=0.60-0.61, D=1.4 [111].

The simulated surface of the material is shown in Fig 5.28 and the N-R relation calculated by variation method is shown in Fig5.15. From Fig 5.29, it is known that the surface of the material is fractal surface, and D_f=1.389, which is very close to the theoretical value *D*=1.4. This result verifies the calculating method and procedure as stated above.

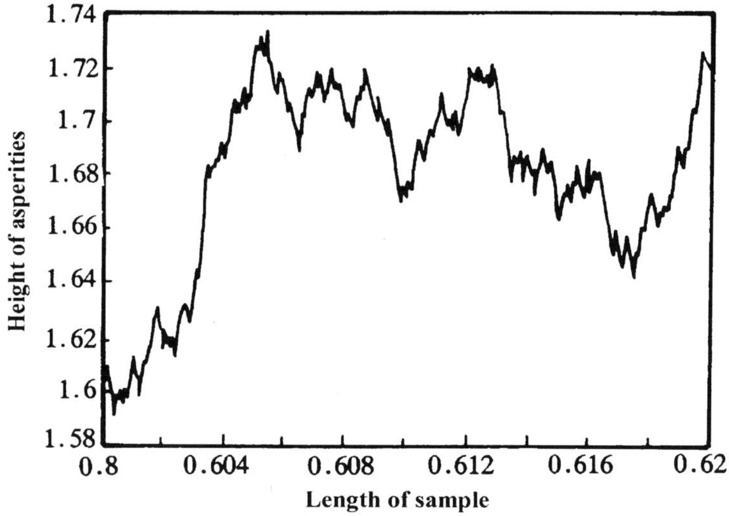

Fig 5.28. Simulated surface of the material

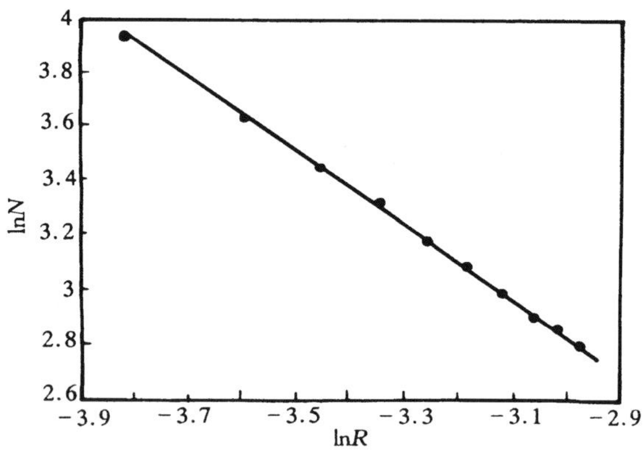

Fig 5.29. N-R relation of the simulated surface of material (D_f=1.389)

5.3.3.3 Determination of the Values of Crack Angle of the Abrasion Patterns

In order to obtain real worn surface of rubber to calculate the value of crack angle, the experimental method as described in Section 2.4.1 was used to perform the abrasion experiments for the SBR and NBR materials. Firstly, the profile of the worn surface of rubber sample was measured by profilometer, and then the fractal dimension of the worn surface was calculated by using the above method. Thus, the value of the crack angle can be determined.

Generally, the worn surfaces display their fractal character only in a specified scale and level. Their upper and lower limits are restricted by a certain characteristic size. This range with self-similarity is called non-scale region. By analyzing the morphology of the worn surface, the pattern width was found in the range of 35 ~ 500μm. Hence, the size of step (measuring scale R) should be chosen also within this range for calculating the measuring number N.

The calculation results of fractal dimensions of the real worn surfaces of rubber are listed in Table 5.1 and shown in Fig 5.30. According to equation (5.69), the corresponding value of crack angle can be determined.

Table 5.1. Calculation results of fractal dimensions of the worn surfaces of two kinds of the rubber

Materials	Normal load N N	Sliding speed v m/s	Frictional distance Tr rev	Fractal dimension D_f	Figures
SBR	10	0.10	150	1.28	5.30,a
SBR	10	0.10	300	1.28	5.30,b
SBR	10	0.10	450	1.31	5.30,c
SBR	8	0.16	1200	1.35	5.30,d
SBR	12	0.16	1000	1.20	5.30,e
NBR	8	0.10	450	1.28	5.30,f
NBR	8	0.10	900	1.28	5.30,g
NBR	8	0.10	1350	1.28	5.30,h
NBR	8	0.16	1800	1.26	5.30,i
NBR	12	0.10	2000	1.13	5.30,j
NBR	5	0.10	2000	1.50	5.30,k

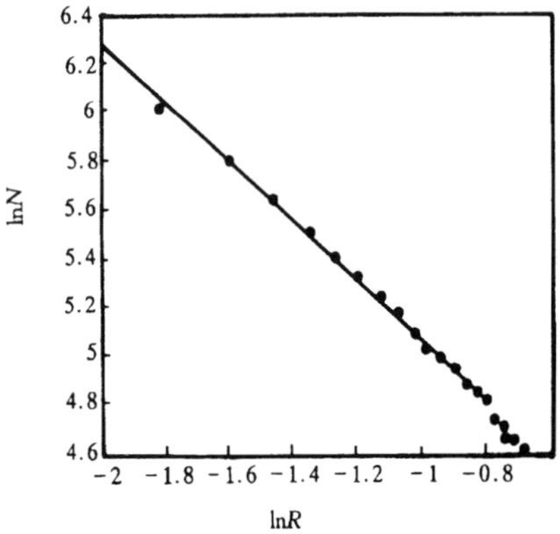

(a) SBR

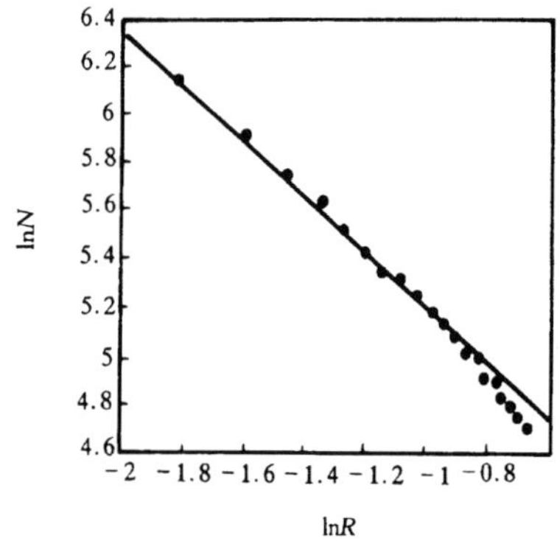

(b) SBR

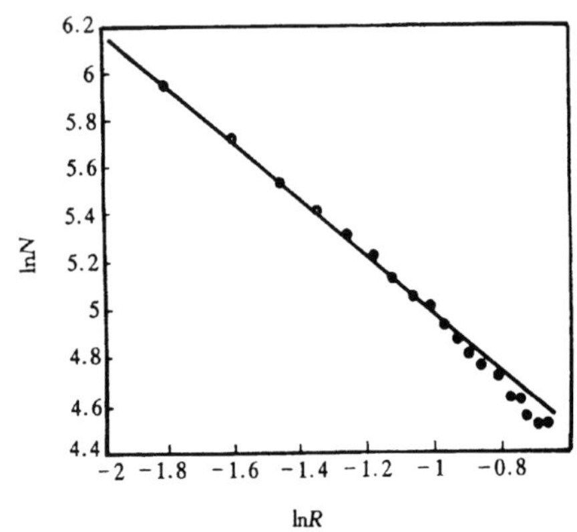

(c) SBR

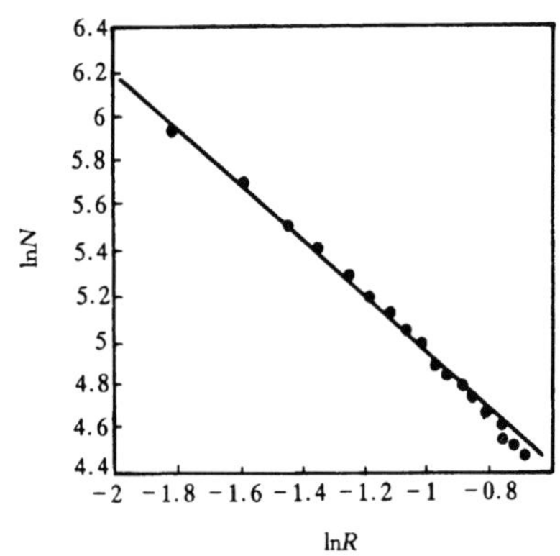

(d) SBR

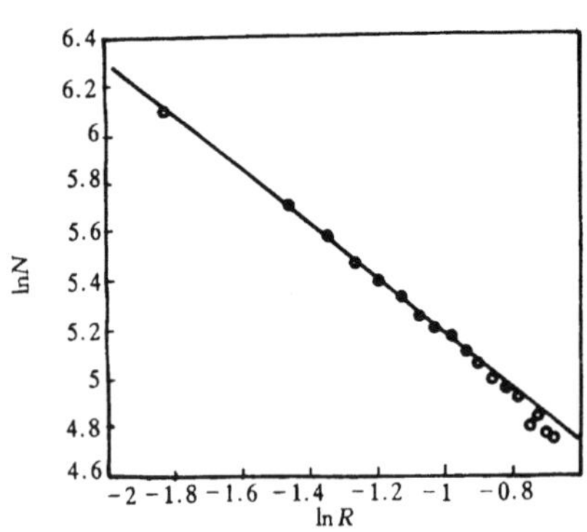

(e) SBR

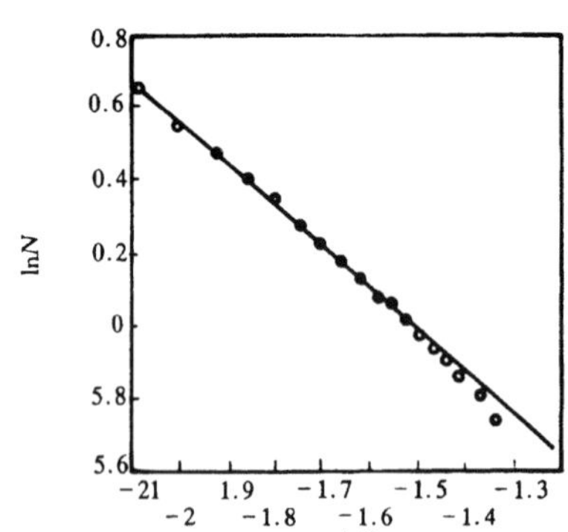

(f) NBR

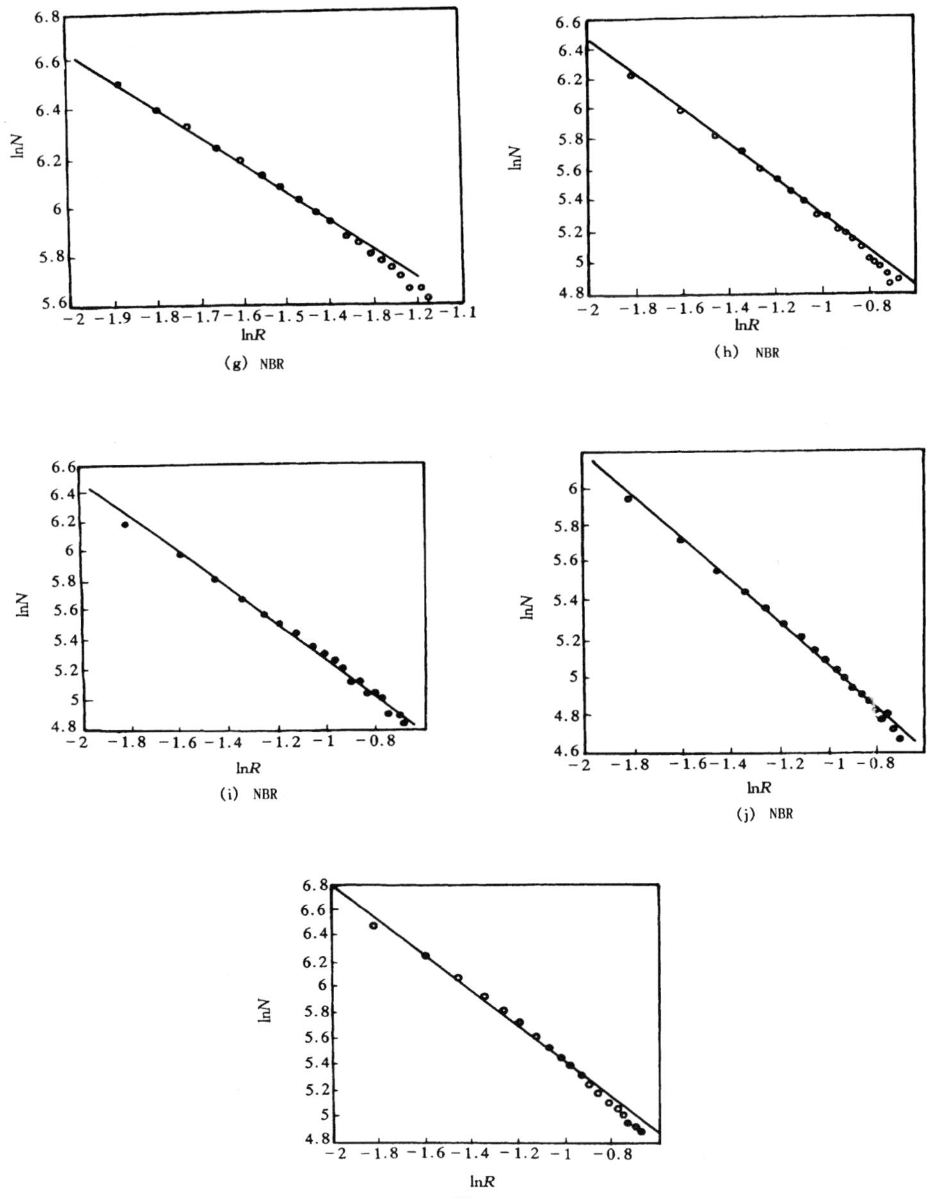

Fig 5.30. Calculation results of fractal dimensions of worn surfaces of SBR and NBR

Following the experimental results obtained by Southern and Thomas [54] and Uchiyama and Ishino [60] respectively, in case of N=10N, the linear wear rate of rubber abrasion by a line contact is of 6μm/rev and the moving speed of the abrasive pattern is of 17μm/rev, hence the corresponding value of crack angle can be calculated, namely =arctan(6/17)=22°. This value is close to the value of crack angle 23° which is determined by the fractal dimension, D= 1.28 based on the calculating method as stated in References [54,60], It demonstrates the present method determined the crack angle is practicable.

5.3.3.4 Factors of Influence on the Fractal Dimension and the Crack Angle of Worn Surface

(1) Frictional distance

In the unsteady state of abrasion, the influence of the frictional distance on the fractal dimension and the crack angle of worn surfaces of SBR and NBR materials is shown in Tables 5.2 and 5.3 respectively (N and v is referred to the normal load and the sliding speed respectively.)

Table 5.2. The variation of fractal dimension and the crack angle of worn surface of SBR with the frictional distance (N=10N, v=0.10m/s)

Frictional distance $T_{f,}$ rev	150	300	450
Fractal dimension D_f	1.28	1.28	1.31
Crack angle α	23°	23°	24°

Table 5.3. The variation of fractal dimension and the crack angle of the worn surface of NBR with the frictional distance (N=8N, v=0.10m/s)

Frictional distance $T_{f.}$ rev	450	900	1350
Fractal dimension D_f	1.28	1.28	1.26
Crack angle α	23°	23°	22°

From Tables 5.2 and 5.3, it can be seen that the fractal dimension of worn surface does not vary with the increase in the frictional distance and so does the corresponding crack angle during the unsteady state of abrasion. The reason is probably that the tongue being torn on the rubber surface is pulled by the frictional force during the wear process, the propagating direction of the crack at the root of the tongue is parallel to the surface of abrasion pattern, thus the corresponding value of crack angle keeps unchangeable.

(2) Normal load

During the steady state of abrasion, the effects of the normal load on the fractal dimension and the crack angle of the worn surface of SBR and NBR materials are shown in Tables 5.4 and 5.5 respectively.

Table 5.4. The variation of fractal dimension and crack angle of worn surface of SBR with the normal load

Normal load N	8	10	12
Fractal dimension D_f	1.35	1.28	1.20
Crack angle α	25°	23°	18°

Table 5.5. The variation of fractal dimension and crack angle of worn surface of NBR with the normal load

Normal load N	5	8	12
Fractal dimension D_f	1.50	1.28	1.13
Crack angle α	33.5°	23°	14°

From Tables 5.4 and 5.5, it is found that the fractal dimension of the worn surface decreases with the increase in normal load and the corresponding crack angle is also decreased during the steady state of abrasion. It is ascribed to that the crack initiation of rubber surface is in accordance with the criterion of maximum shear stress [19]. The shear stress of rubber surface increases with the increase in normal load, and the angle included between the iso-(shear)stress line and the horizontal line is also increased with the rise in the shear stress [19]. Therefore, for same material, the angle included between the initiating and propagating direction of crack and the horizontal line decreases with the increase in normal load, then the corresponding fractal dimension is also decreased. Owing to the fractal dimension keeps constant in the unsteady state of abrasion, the fractal dimension during the steady state of abrasion decreases with the increase in normal and so does the corresponding crack angle.

5.3.4 Computer Simulation of Worn Surface of Rubber Abrasion

On the basis of the fractal analysis of rubber surface, a computerized simulation of the worn surface of rubber abrasion can be made according to the fractal theory and wear mechanism, thus checking and deepening the understanding of wear mechanism by the analysis on the simulated worn surface of the rubber.

In light of the experimental observation of dry abrasion by a line contact as stated in Section 4.1.2, and using the generating element model of worn surface as shown in Fig 5.25, the overlap levels of abrasion pattern is classified as three grades from large to small in size in order to simulate the worn surface much better (Figure 5.31).

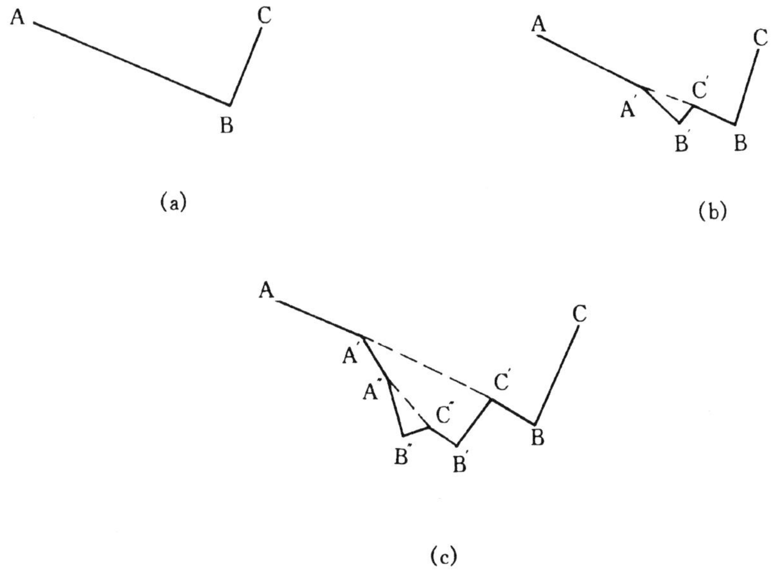

Fig.5.31. Scheme of the overlap levels of abrasion pattern
(a) primary abrasion pattern; (b)secondary abrasion pattern;
(c) third-level abrasion pattern

As stated previously, the spacing of the primary abrasion patterns of worn surface of rubber is approximately even in the steady state of abrasion. Hence the length of line AC might be regarded as unchangeable roughly.

According to the results of fractal analysis of the worn surfaces of rubber as mentioned previously (Fig 5.30), the worn surface is fractal. Therefore, this fractal surface must have overlap levels. It means that a secondary abrasion pattern must be inlaid the primary abrasion pattern, namely, within line AC (Fig 5.31, b) and the length of line A′C′ is random. Moreover, the third-level abrasion pattern possibly also exists in the secondary abrasion pattern (Fig 5.31, c), in which, the length of line A′′C′′ is also random.

The simulation of the worn surface of rubber has been realized on the computer, and the corresponding computing flowchart is shown in Fig 5.32. According to the measuring results of the worn surface, the value of crack angle is generally about 22°, therefore, the value of crack angle is chosen as 22° for computerized simulation. The simulation results are shown in Fig 5.33, in which, three levels of the abrasion patterns are observed on the worn surface of rubber.

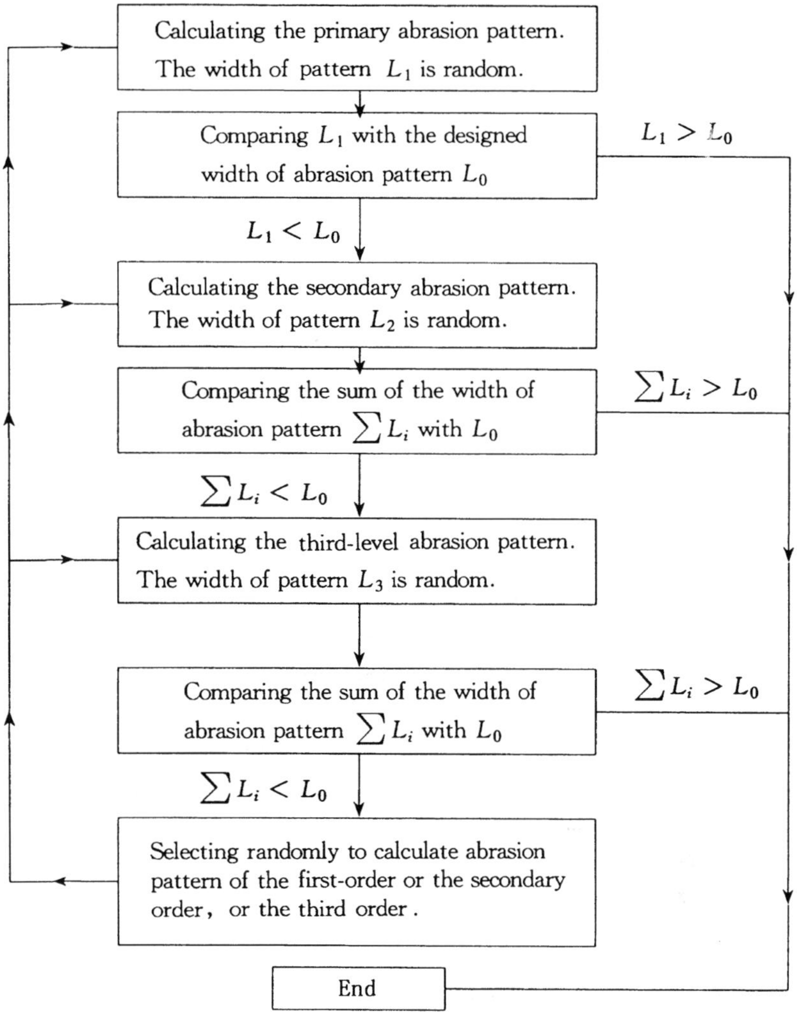

Fig 5.32. Computing flowchart of abrasion pattern of rubber

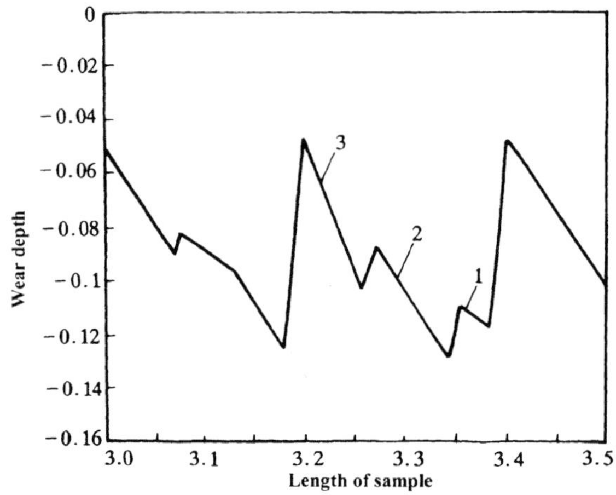

Fig 5.33. Computerized simulation of worn surface of the rubber
1-primary abrasion pattern; 2-secondary abrasion pattern;
3-third-level abrasion pattern

The dimension of the simulated worn surface is further calculated by using the variation method, and the R-N relation obtained is shown in Fig 5.34. The calculated dimension, D_f is equal to 1.28, which is close to the theoretical value 1.26. This result verifies the above simulation method of abrasion pattern of the rubber and the theory of rubber abrasion by a line contact as stated in Section 5.1.

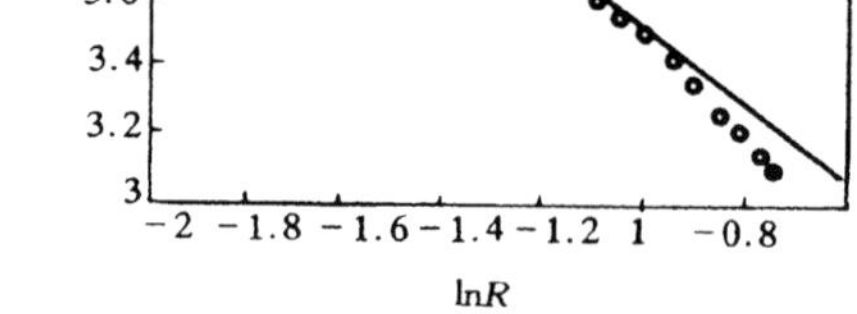

Fig.5.34. R-N relation of simulated worn surface of the rubber

5.4 Computer Image Analysis of Wear Debris

Debris, as the last remains of abrasion process, is the comprehensive result of material suffering a series of mechanical, physical and chemical actions. Therefore, study in the shape and its distribution of debris is of great importance to reveal the rules of occurrence and growth of wear. Furthermore, the running condition of the machine can be monitored and its wearing locations can be judged according to the type, shape, quantity and particle size of debris during the running process of machine. Therefore, the ferrography technology for separating and analyzing the wear debris and pieces of metal from the lubricating oil, which was developed in 1980's, has played an important role on judging the type and degree of wear of components and monitoring the behaviors of machines. However, this analysis technique of wear debris is unapplicable for elastomers. For this reason, a method of image analysis with computer for the analysis of wear debris of rubber is presented here.

5.4.1 Basic Principle

5.4.1.1 Collection and Treatment of Debris Image

During the running process of machine, the debris is collected in specified time, and then the debris image is obtained with microscope and recorded in computer with image-recorder. The debris boundary is defined by gradient method of the technique of treatment and analysis of image, thus the amount, projected area and perimeter of debris are obtained. Finally the shape factor G, which is the ratio of the equivalent projected area $\boldsymbol{S}$ to the real measured one A, can be acquired by the following equation:

$$G = S / A \tag{5.81}$$

In order to determine the shape factor G, the equivalent diameter of each debris should be firstly calculated by the following equation with the measured perimeter L and the measuring projected area A of debris:

$$D = 4A / L \tag{5.82}$$

Thus the equivalent projected area of image (debris) S can be obtained by the following equation:

$$S = \frac{\pi D^2}{4} = \frac{4\pi A^2}{L^2} \tag{5.83}$$

According to the definition of the shape factor G, substituting the above equation into equation (5.81), we have:

$$G = \frac{4\pi A}{L^2} \tag{5.84}$$

The more the shape factor approaches unity the more similar to a ball the shape of debris is. Since the value of shape factor is inversely proportional to the square of perimeter, it might reflect the smooth degree of the debris boundary, thus reveal the formation mechanism of debris to a certain extent.

5.4.1.2 Debris Analysis

It is assumed that S is the average projected area of debris in steady state of abrasion, N_1 and N_2 are the numbers of debris whose projected area is more or less than S respectively. Since N_1 is obviously more than N_2 when the irregular wear happens, the parameter (N_1-N_2) can be regarded as the criterion to judge if accident wear occurs. Parameter (N_1+N_2) is the number of debris appeared per unit time. Assuming G' to be the average shape factor of debris during the steady state of abrasion and G being the average shape factor of debris in the process of abrasion, parameter G/G' shows the variety of shape of debris. As the wear rate will increase obviously and the shape and size of the debris will also change much evidently on the eve of the machine in trouble, the index of abradibility *I* can be set up as

$$I = (N_1 + N_2)(N_1 - N_2)(G/G') \tag{5.85}$$

Therefore, the process and degree of wear of material can be analyzed on the basis of the curve of the index of abradibility *I* varying with the time.

5.4.2 Examples

The SBR and NBR samples were tested according to the experimental method as described in Section 2.4.1, under the condition of a constant sliding speed of 0.10m/s, corresponding to n=50r/min, in order to obtain debris in different normal loads and wear stages. And then the images of debris were collected in perspective light with the microscope Leica Mps 48. The fixed magnification was chosen to be 50 as the size of debris is relative large. With images recorder, the images of debris under different experimental conditions were input into computer. The value of gradient of each image element of digital image was taken into accounted, and then a proper threshold value was defined, because the values of grey scale of image element of both sides of debris border are obviously different. The image element points, whose gradient values of grey scale are larger than the threshold value, are the points of step border. Thus the border of debris can be defined by the gradient method. Consequently, the perimeter of debris border and the projected area were obtained.

5.4.2.1 Distribution Function of the Shape Factor of Debris

According to the calculating results, the relative frequency rectangular plots of shape factor of debris of SBR and NBR materials under different experimental conditions and different stages of abrasion were obtained. Fig 5.35 illustrates the relative frequency rectangular plots of shape factor of debris. As roughly judged by the above rectangular plot, the distribution of shape factor belongs to a normal distribution and can be generally described as follows:

$$F(G) = \frac{1}{(2\pi)^{1/2}\sigma}\int_0^G \frac{e^{-(t-\mu)^2}}{2\sigma^2}dt \tag{5.86}$$

Where, μ- average value of shape factor;

σ - average variance of shape factor;

t - microbody of shape factor;

The value of σ and μ can be determined by the method of maximum likelihood estimation, and assuming:

$$H_0{:}F \backsim G(\mu,\sigma^2) \tag{5.87}$$

After transferring the above equation into standard normal distribution, the normal testing was done by applying the Pearson's x^2 estimation. Comparing the value obtained in the x^2 testing table with that value in the x^2 distribution table, it is shown that H_o is accepted under the level 0.1, which means the data being normal distribution.

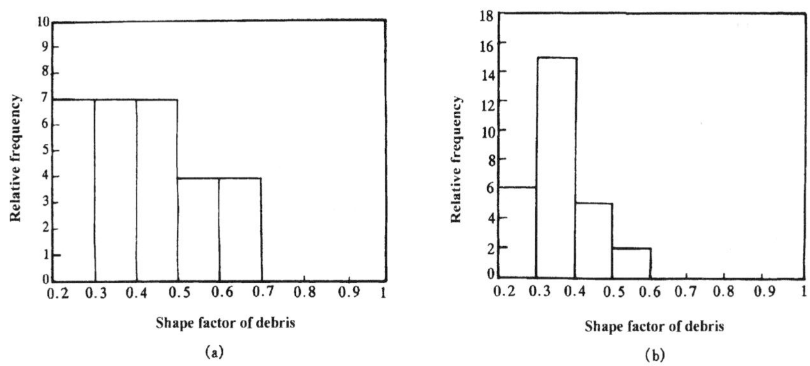

Fig. 5.35. Relative frequency rectangular plots of shape factor of rubber debris (a) SBR (N=12N, T_r=450r); (b) NBR (N=8N, T_r=1350r)

On the basis of the results of experiment and analysis, the rules of the shape factor of debris of SBR and NBR materials varied with the frictional distance T_r during the unsteady state of abrasion were obtained as shown in Tables 5.6 and 5.7 respectively. In the tables, N is normal load.

Table 5.6. The average value μ and variance σ of shape factor of debris of SBR varied with the frictional distance during the unsteady stage of abrasion (N=10N)

Tr, r	150	300	450
μ	0.375	0.51	0.47
σ	0.14	0.09	0.24

Table 5.7. The average value μ and variance σ of shape factor of debris of NBR varied with the frictional distance during the unsteady stage of abrasion (N=8N)

Tr, r	450	900	1350	1800
μ	0.39	0.38	0.31	0.34
σ	0.15	0.16	0.08	0.09

The changing rule of the debris shape factor of SBR and NBR materials varied with the normal load during the steady stage of abrasion is shown in Table 5.8 and Table 5.9 respectively.

Table 5.8. The average value μ and variance σ of shape factor of debris of SBR varied with the normal load during the steady stage of abrasion (T_r=480r)

N, N	5	8	10	12
μ	0.38	0.40	0.47	0.37
σ	0.10	0.12	0.24	0.13

Table 5.9. The average value μ and variance σ of shape factor of debris of NBR varied with the normal load during the steady stage of abrasion (T_r=1800r)

N, N	5	8	10
μ	0.36	0.34	0.33
σ	0.14	0.09	0.16

From Tables 5.6, 5.7, 5.8 and 5.9, it can be seen that the values of shape factor of debris for two kinds of rubber (including the unsteady and steady stage) are all in the range from 0.3 to 0.5, which shows that the debris is mostly strip, as consistent with the SEM observation. The rupturing length of tongue of the debris is far less than its width. It indicates that the rupture of tongue is fatigue fracture, thus proves that the physical model of abrasion process being set up in the fatigue – fracture theory of dry abrasion by a line contact as mentioned in Section 5.1 is correct.

5.4.2.2 Distribution of Area of Debris

According to the measuring results, the relative frequency rectangular plots of projected area of debris of SBR and NBR materials during the unsteady state of abrasion were gained (Fig.5.36). It can be deduced preliminarily from the plots that the area distribution of debris is roughly similar to the normal distribution and able to be characterized as

$$F(G') = \frac{1}{(2\pi)^{1/2}\sigma'} \int_0^{G'} e^{-2(t'-\mu')} / 2\sigma'^2 dt' \tag{5.88}$$

Where, μ' - average value of projected area;

σ'- average variance of projected area;

t'- microbody of projected area.

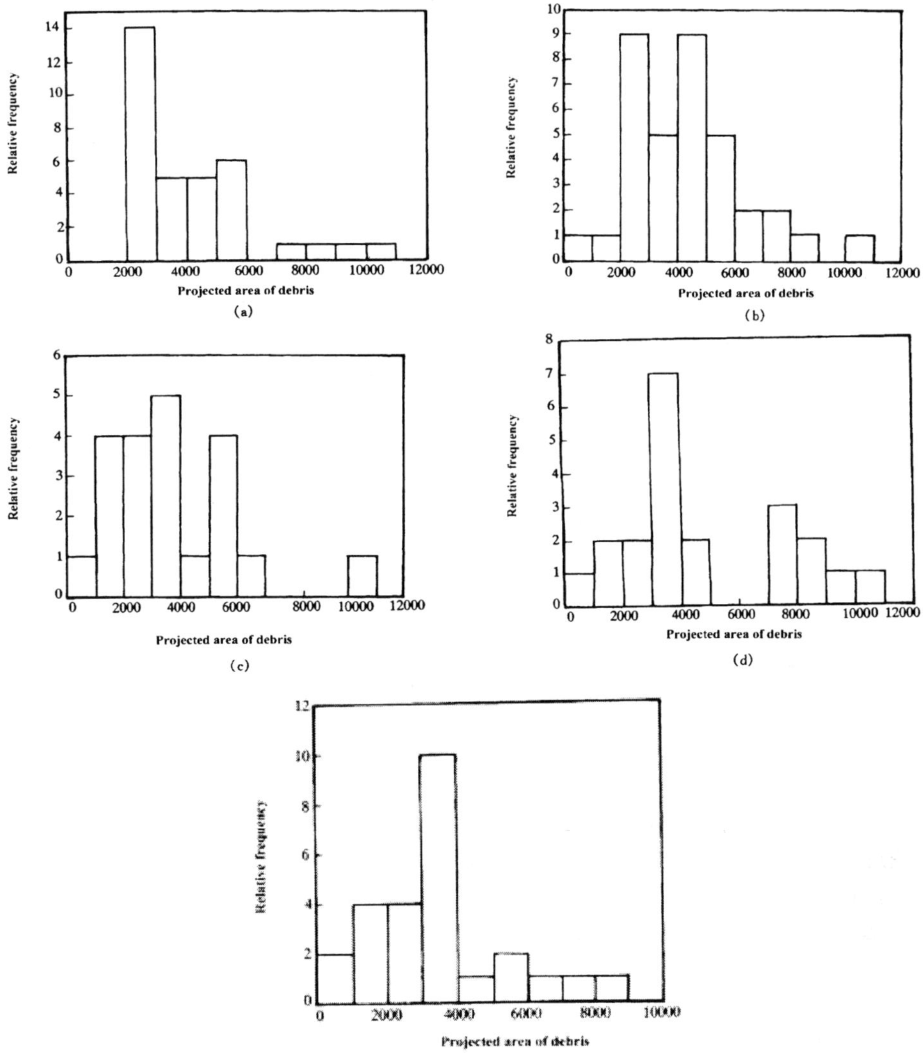

Fig. 5.36. Relative frequency rectangular plots of the projected area of debris during the unsteady state of rubber abrasion

(a) SBR (N=10N, Tr=150r); (b) SBR (N=10N, Tr=300r); (c) NBR (N=8N, Tr=450r); (d) NBR (N=8N, Tr=900r); (e) NBR (N=8N, Tr=1350r)

The average value μ' and average variance σ' can be acquired by the method of maximum likelihood estimation, and assuming

$$H_0 : F \backsim G'(\mu', \sigma'^2) \tag{5.89}$$

After transferring the above equation into standard normal distribution, the normal testing was done by applying the Pearson's x^2 estimation. Comparing the value gained in the x^2 testing table with the value in the x^2 distribution table, it is shown that H_0 is accepted under the level 0.1, which means the data being normal distribution.

According to the results of experiment and analysis, the rules of the projected area of debris of SBR and NBR materials varied with the frictional distance T_r during the unsteady state of abrasion were achieved, as listed in Tables 5.10 and 5.11 respectively.

Table 5.10. The average value μ' and variance σ' of projected area of debris of SBR varied with the frictional distance during the unsteady state of abrasion (N=10N)

Tr, r	150	300	450
μ'	3421	3837	3379
σ'	2114	2069	2427

Table 5.11. The average value μ' and variance σ' of projected area of debris of NBR varied with the frictional distance during the unsteady state of abrasion (N=8N)

Tr, r	450	900	1350
μ'	3500	4782	2960
σ'	2589	3152	1933

From Tables 5.10 and 5.11, it is shown that the projected area of debris is unchangeable approximately during the unsteady state of abrasion.

Based on the measuring result, the relative frequency rectangular plots of projected area of debris of SBR and NBR materials during the steady state of abrasion under the action of different normal loads were obtained (Figure 5.37). It could be deducted preliminarily from the plots that the distribution of debris area is similar to uniform distribution, and the corresponding density function is expressed as

$$\psi(x) = \begin{cases} 1/(b-a) & a < x < b \\ 0 & \text{the others} \end{cases} \tag{5.90}$$

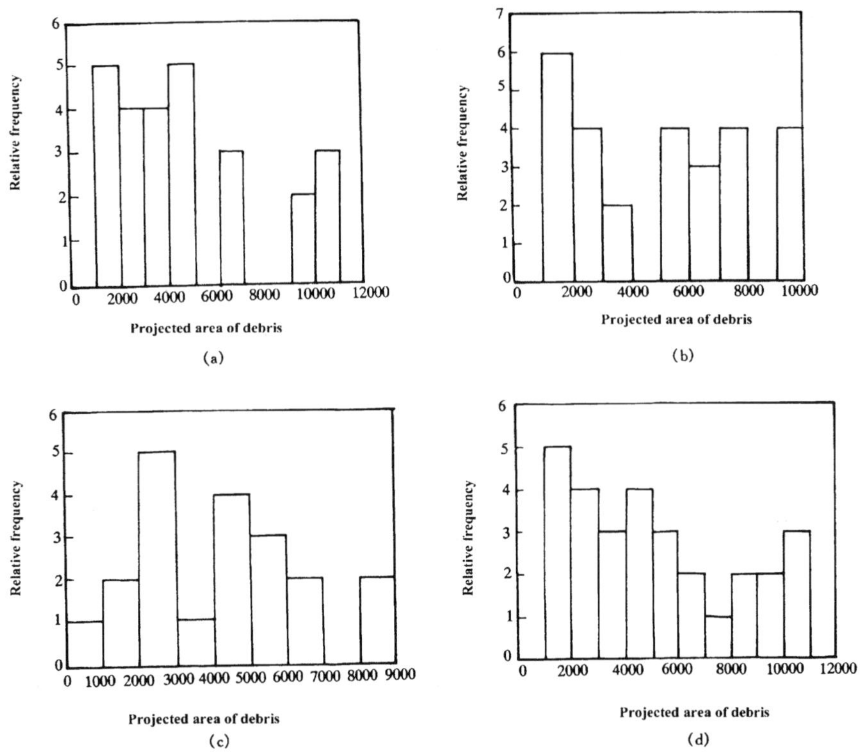

Fig. 5.37. Relative frequency rectangular plots of the projected area of debris during the steady state of rubber abrasion
(a) SBR (N=5N, Tr=450r); (b) SBR (N=12N, Tr=450r); (c) NBR (N=8N, Tr=1800r); (d) NBR (N=10N, Tr=1800r)

Because the distribution of the projected area of debris is uniform under the action of different normal loads, it might be deduced that the volume distribution is also even roughly. As the generation of debris is mainly resulted from the fatigue rupture, the number of dynamic stress cycle n which cause the debris to be peeled off, can be determined by equation (7.1) in *chapter 7* based on the fatigue theory of wear [114].

It is known from the equation (7.1) that the larger the normal load, the more the value of repeated dynamic stress is, and the less the value of n is, thus the more the debris generated per unit time will be. Consequently, the wear rate during the steady state of abrasion is increased with the increase in normal load that is accord with the experimental results. Thus it can be seen that the area distribution of debris during the steady state of abrasion verifies the dropout mechanism of debris of the fatigue-fracture theory of rubber abrasion by a line contact as stated in Section 5.1 once again.

Chapter 6

EROSION

Erosive wear or erosion is defined as the wear resulting from the interaction between a solid surface and a fluid stream containing abrasive particles with a certain speed, or the impact of free-moving liquid (or solid) particles on a solid surface. When the erosion is caused by the fluid stream flowing nearly parallel to or being parallelized the solid surface, it is called abrasive erosion. When the erosion occurs under the condition of the fluid stream flowing approximately normal to or at right angles to the surface of solid, it is termed impact erosion. The values of erosion are usually expressed as the gravimetric rate of erosion, namely gravimetric eroded rate:

$$\varepsilon = \frac{\text{loss in weight of material}}{\text{weight of abrasive particles}}$$

Thus the volumetric rate of erosion or volumetric eroded rate is given by

$$\varepsilon_v = \varepsilon / \rho \tag{6.1}$$

Where ρ is the density of materials.

6.1 Mechanisms of Abrasive Erosion

Abrasive erosion is such a kind of erosion that its angle of impact is approximate or equal to zero. It is also termed sliding erosion by some scholars [115]. This kind of wear usually occurs in various fluid machines and the corresponding pipelines to transport the fluid medium containing solid particles. It is one of the common types of wear of materials used in petroleum, coal and nonferrous metal industries and so on. However, the study of abrasive erosion being regarded as an independent type of erosion did not conduct extensively. In recent years, we have systematically and successively investigated the mechanisms of abrasive erosion of NR, SBR, NBR, polyurethane(PU), and fluoroelastomer(FE) in several fluid media containing quartz particles (water, polyacrylamide solution and sodium hydroxide solution) using two types of abrasive erosion test machines designed by ourselves [116~118]. The main research results are discussed in this section.

6.1.1 Experimental Procedure [116]

Experiments were conducted using such an abrasive erosion testing machine that was designed according to the basic characteristics of abrasion erosion (Figure 6.1). The fluid medium containing solid grains flows from the slurry tank 10 into the slurry pump 2, and then runs up the test cavity 7 and back to the slurry tank. The rubber sample is fixed in the test cavity as shown in Figure 6.2.

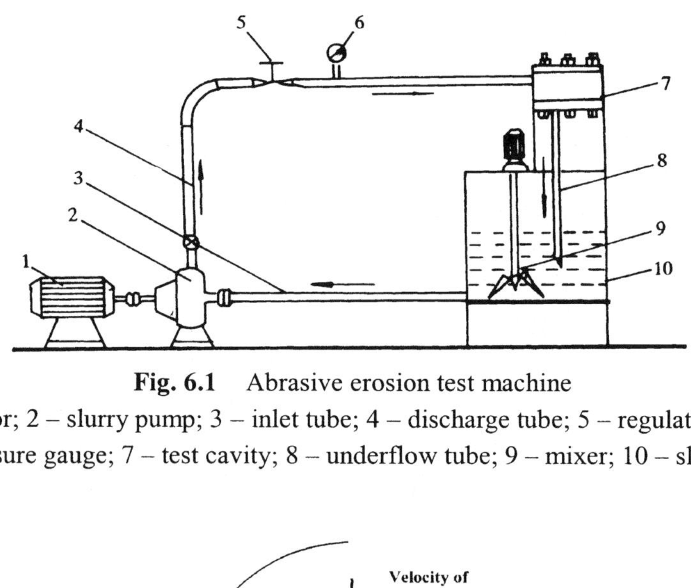

Fig. 6.1 Abrasive erosion test machine

1 – motor; 2 – slurry pump; 3 – inlet tube; 4 – discharge tube; 5 – regulating valve; 6 – pressure gauge; 7 – test cavity; 8 – underflow tube; 9 – mixer; 10 – slurry tank;

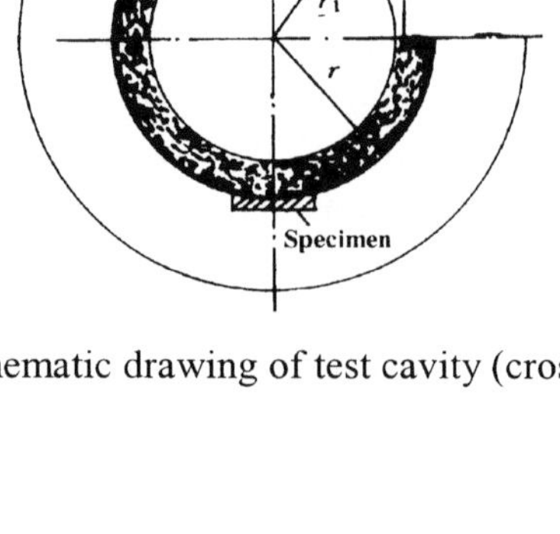

Fig.6.2 Schematic drawing of test cavity (cross section)

The main testing parameters are given as follows: velocity of flow, 7.2~11.1m/s; particle concentration of medium, 5~30wt%; size of quartz particle, 76~450 μm; Mohshardness of particle, 7.

6.1.2 Morphological Characteristics of the Worn Surfaces and the Physical Processes of Abrasive Erosion

6.1.2.1 Polyurethane (PU)

Some traces of impact, indentation and scratch were found on the eroded surface in different media by the SEM examination. A large number of cracks and scrap peeling were observed on the eroded surface in water or polyacrylamide (PAM) solution. However, only the impacting pits, scratch traces and plastic accumulations of materials were found on the eroded surface in sodium hydroxide (NaOH) solution. The morphology of the eroded surfaces of PU material in water containing quartz particles, PAM and NaOH solutions are shown in Figures 6.3, 6.4 and 6.5 [116,117] respectively. Most of the debris is in scraps form.

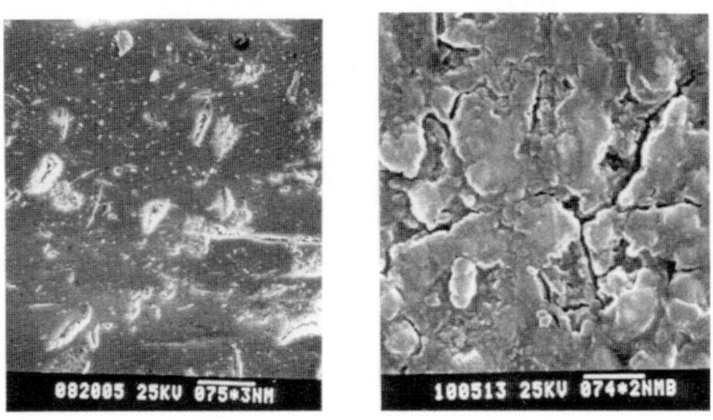

(a) (b)

Fig.6.3 Morphology of the worn surface of polyurethane (in water)
(a) indentations and scratches; (b) cracks and delaminations

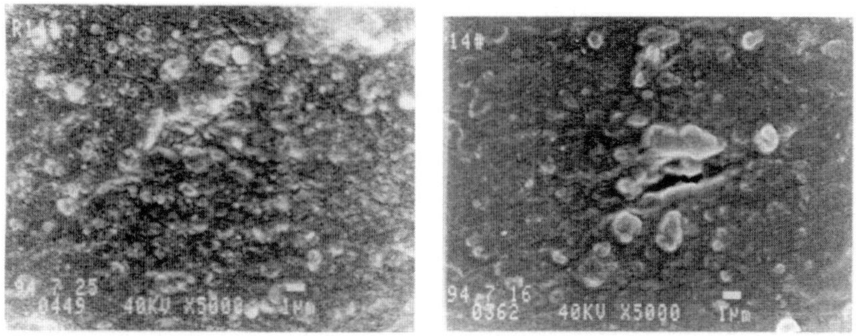

(a) (b)

Fig.6.4 Morphology of the worn surface of polyurethane (in PAM solution)
(a) impact indentations; (b) cracks and peel off

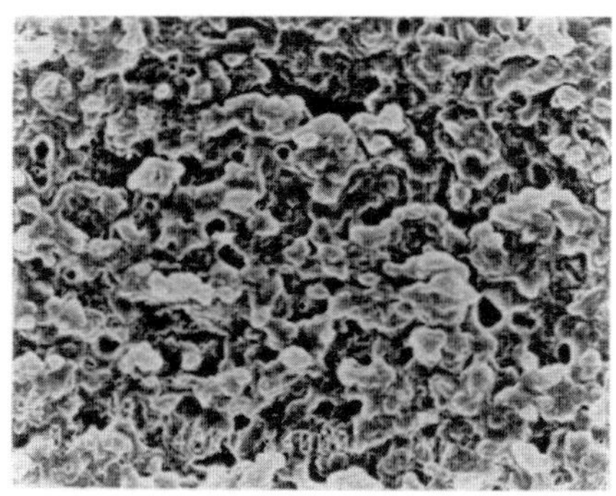

Fig.6.5 Morphology of the worn surface of polyurethane (in NaOH solution)

Obviously, microtearing and microdeforming occur on the eroded surface by the action of microcutting of the flowing abrasive particles with a certain amount of kinetic energy, which cause the surface and subsurface to generate tensile, compressive and shearing stresses. Consequently, under the continuous and multiple acting of the flowing abrasive grains, crack growth appears in the surface or subsurface layer of PU material. The crack nucleation, either is possibly the faults in material, such as voids, impurities, etc., or is resulted from the fracture of molecular chains or the intermolecular bonds induced by the mechanical action or thermal effect. The fracture of molecular chains can produce the active free-radicals. Thus the radicals transfer to the adjacent molecular chains, and result in the fracture of molecular chains again. As a result, the crack is gradually formed (for further details, see Chapter 8).

Being similar to the metal, the germinating and propagating of cracks are always generated at a certain depth beneath the surface or subsurface layer being parallel with the surface since the maximum shear stress usually occurs at that depth. Moreover, this depth will gradually decrease with the increase in the tangential force (frictional force). In addition, quite a number of heat accumulations induced by the friction and deformation hysteresis of PU material usually lead to that the temperature of the substrate is higher than that of the surface because the heat-sinking condition of the surface is better than that of the substrate. As s result, it could change the orientation of the molecular chains and result in the decrease in the cohesive energy of the material between the surface layer and substrate. Where the higher temperature is, the lower the cohesive energy is. Therefore, the cracks are initiated firstly at a certain depth beneath the surface layer of material.

Since the abrasive particles flow continuously and act on the solid surface many times, thus the crack growth could appear along with the tip of crack and extend finally through to the solid surface under the action of repeated stress. Consequently, the surface layer of material is delaminated in debris form (Fig.6.3). Thus it can be deduced that the mechanism of abrasive erosion of PU material is mainly fatigue delamination (high-frequency fatigue) of material resulted from the microcutting (ploughing) and plastic deformation. The fatigue is usually able to accelerate the aging of rubber. Conversely, the aged rubber will also intensify the fatigue wear. The physical model of the process of abrasive erosion is illustrated in Figure 6.6.

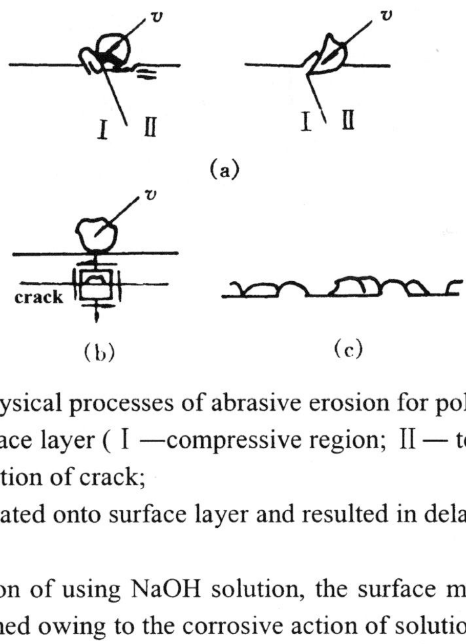

Fig.6.6 Physical processes of abrasive erosion for polyurethane
(a) microcutting in surface layer (I —compressive region; II — tensile region);
(b) growth and propagation of crack;
(c) cracks being propagated onto surface layer and resulted in delamination of materials

Under the condition of using NaOH solution, the surface mechanical properties of PU material are weakened owing to the corrosive action of solution. Thus, the material is ploughed easier, and worn more intensive. However, no obvious fatigue delamination was observed.

6.1.2.2 Styrene-Butadiene Rubber (SBR)

A sticky layer and a number of indents observed on the abraded surface caused by the eroding of flowing water containing quartz particles (Fig. 6.7). In front of the indents, the tongues to be peeling off (Fig.6.7(a)) and some ridges like lip along with the flowing direction of fluid (Fig.6.7(b)) were found. The debris is almost taken the shape of crimp.

(a)

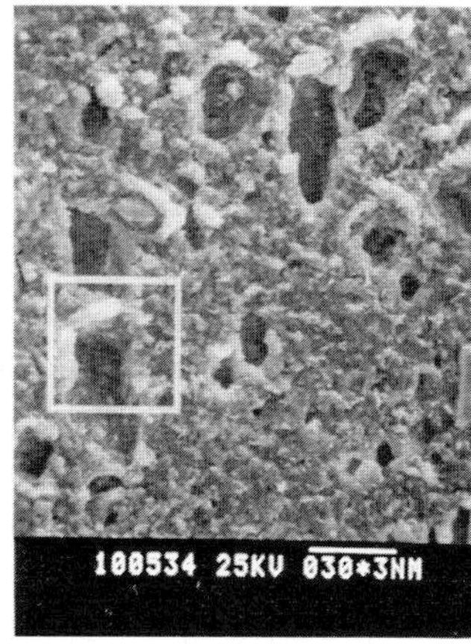

(b)

Fig. 6.7 Morphology of the worn surface of SBR (in water)
(a) sticky layer; (b) indents and ridges

Similar to the process of abrasive erosion of polyurethane, microdeforming and microtearing of the surface layer were resulted from the microcutting of the particles possessed a certain amount of kinetic energy. However, these mechanical effects are more severe than that for PU material because the strength and hardness of SBR materials are lower than that of PU material, specially, its tear resistance is much lower. Consequently, under the action of microcutting, the microtearing resulted in the rupture of macromolecules, and then various reactive radical species were formed. On the other hand, owing to the high elastic hysteresis property of SBR material, a large amount of heat might be accumulated in the surface and subsurface layers due to the intense microdeformation, which caused its chemical bonds to be weakened and then produced thermal decomposition. As a result of the rupture of molecular chins and thermal degradation, a sticky layer was formed on the rubber surface [50]. When the sticky layer is worn down or worn off, the surface of rubber will be acted again by the abrasive particles, and then the microdeforming and microtearing will ensue on the rubber surface. As the above processes occur repeatedly, the surface layer of SBR materials will be worn off in the way of microcutting and degradation resulted from the mechanical action, the physical model of this process can be set up as shown in Figure 6.8.

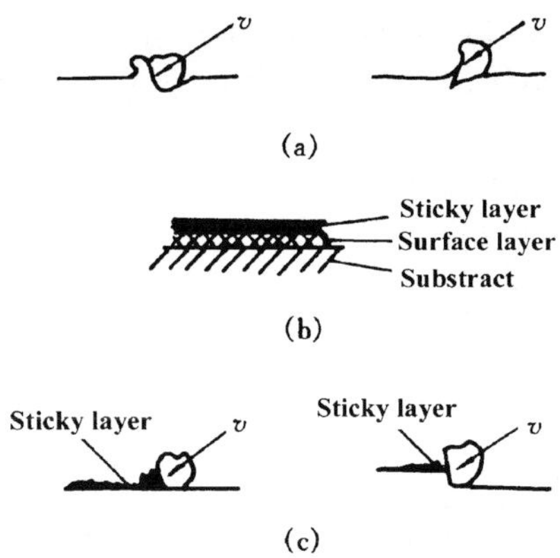

Fig.6.8 Physical processes of abrasive erosion for SBR materials
(a) microcutting in surface layer;
(b) sticky layer formed on the surface of substrate;
(c) microcutting in sticky layer

Although both SBR and PU materials are in the rubbery state, the frictional coefficient and the tangential force (frictional force) of SBR are larger than that of PU under the acting of particles, thus the site of its maximum shear stress is more closed to the surface of rubber. Moreover, SBR material is filled with carbon black. One of the effects of carbon black is improved the thermal conductivity and heat resistance of material, which can lessen the amount of reduction in cohesive energy between the surface layer and the substrate of rubber caused by high temperature. Therefore, the SBR material does not suffer from fatigue delamination as PU material does during the process of abrasive erosion.

6.1.2.3 Natural Rubber (NR)

A sticky layer and ridges like villus were found on the eroded surface under the condition of the abrasive erosion in the media of water, PAM and NaOH solutions as shown in Figures 6.9, 6.10 and 6.11 respectively, which must be accounted for that the microcutting action of the abrasive grains resulted in the rupture of molecular chains, and then the partial ruptured molecular chains produced chemical reaction with the substrate and formed branch chain connection on the surface layer.

Fig. 6.9 Morphology of the worn surface of NR (in water)

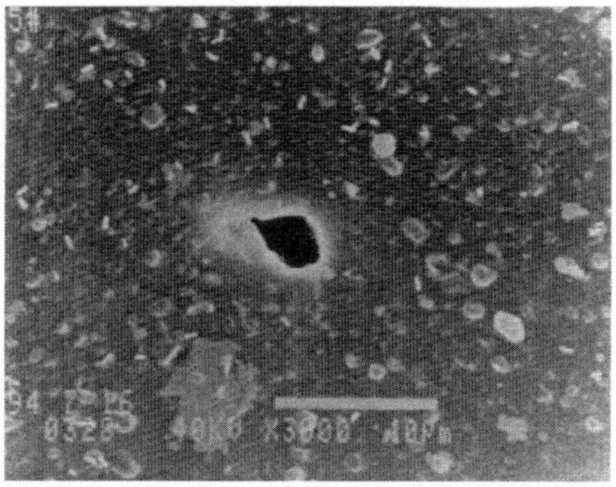

Fig. 6.10 Morphology of the worn surface of NR (in PAM solution)

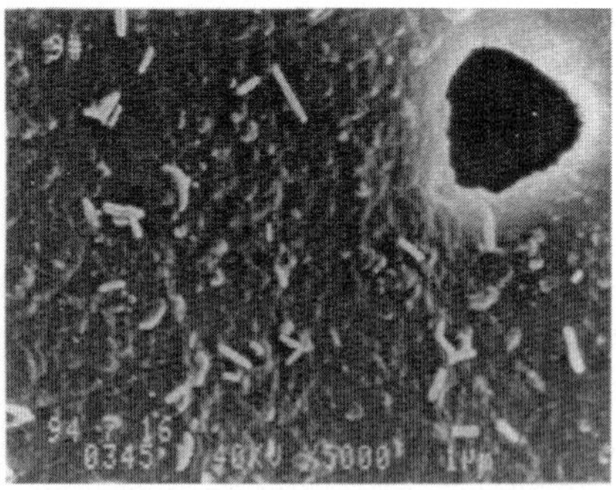

Fig. 6.11 Morphology of the worn surface of NR (in NaOH solution)

Some pits produced by the mechanical action were observed on the worn surface caused by abrasive erosion in the medium of water. At the edge of pits, the plastic accumulations were found. However, on the eroded surfaces resulted from the abrasive erosion in the media of PAM and NaOH solutions respectively, only corrosive pits were found.

Since the mechanical property of NR material is similar to that of the SBR material, the physical process of abrasive erosion of natural rubber in the medium of water is much the same as that of the SBR material, namely, the mechanisms of abrasive erosion are mainly microcutting and decomposition caused by mechanical action. However, the difference between them is that the molecular chain flexibility of NB material is better, moreover, the carbon-carbon bonds on the molecular chain is also much more and is oxidized easier. In addition, in the PAM and NaOH solutions, the corrosive pits were occurred on the surface of rubber due to the corrosive action of the solutions.

6.1.2.4. Nitrile Rubber (NBR)

Some pits and the plastic-broken pieces produced by the mechanical action were found on the eroded surface of the NBR material in all the three media: water, PAM and NaOH solutions, as shown in Figures 6.12, 6.13 and 6.14. Moreover, the cracks and layer delaminated from the surface layer also observed on the worn surface under the condition of abrasive erosion in water or PAM solution.

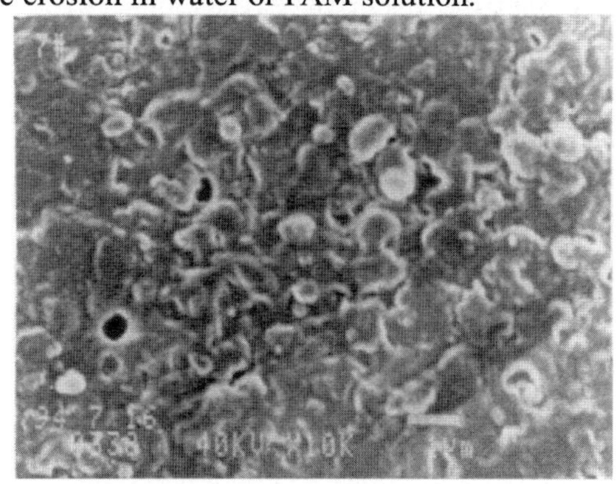

Fig. 6.12 Morphology of the worn surface of NBR (in water)

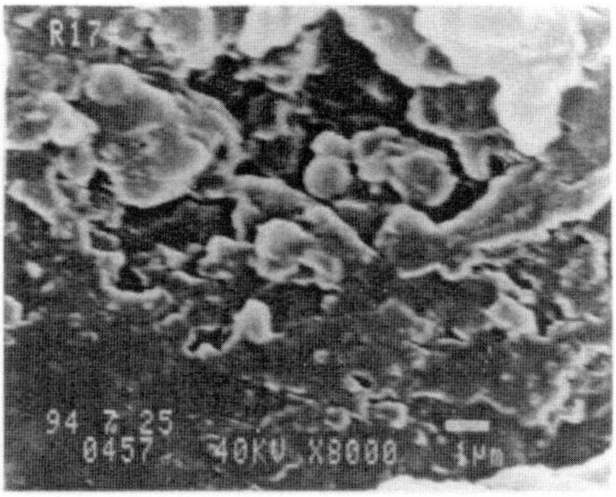

Fig. 6.13 Morphology of the worn surface of NBR (in PAM solution)

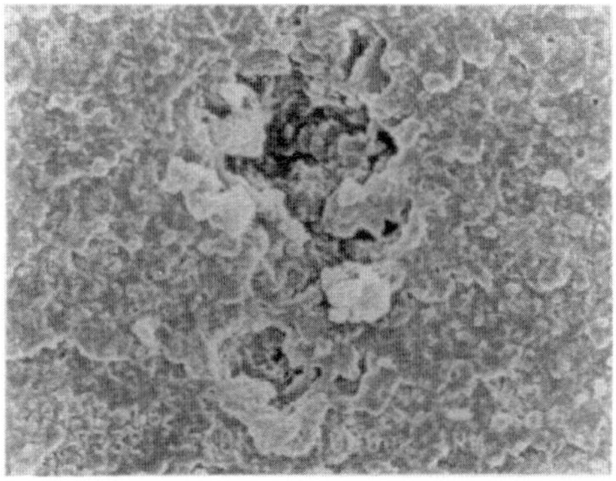

Fig. 6.14 Morphology of the worn surface of NBR (in NaOH solution)

According to the primary characteristics of the eroded surface and in consideration of that the nitrile rubber is about the same as the polyurethane in mechanical properties, it could be deduced that both the materials have similar wear process, i.e., fatigue delamination (high frequency fatigue), but the NBR material has much better heat-sinking performance owing to it contains carbon black. Obviously, the corrosive action of medium can also intensify the above physical process of wear.

6.1.2.5. Fluoroelastomer (FE)

A stretch of fatigue delaminated region was found on the worn surface of fluoroelastomer eroded in water containing quartz particles (Fig. 6.15). From this we may infer that the wear process of fluoroelastomer is similar to that of PU material. However, as the fluoroelastomer does not contain carbon black, its thermal conductivity is rather poor. Therefore, the fracture of molecular chains and thermal decomposition also occurred at the site of subsurface layer where the maximum shear stress and the highest temperature produced, which resulted in the cracking and fatigue delaminating.

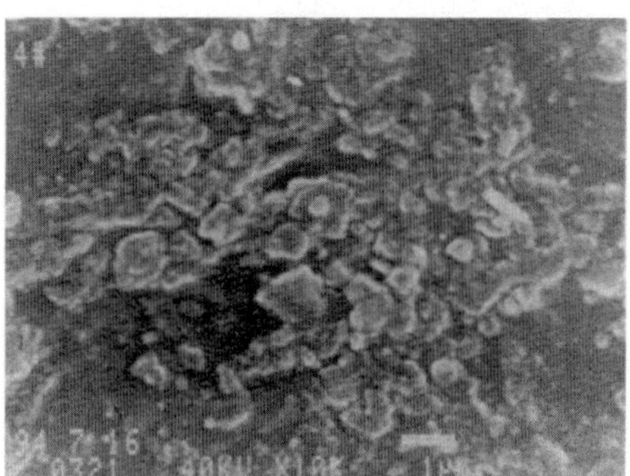

Fig. 6.15 Morphology of the worn surface of FE (in water)

In the media of PAM and NaOH solutions containing quartz grains, the plastic-broken cracks and the platelike debris being about to peel off were found on the eroded surface of fluoroelastomer (Figures 6.16 and 6.17). It is much resembled PTFE in the wearing characters of abrasive erosion in water containing quartz particles [116]. As the fluoroelastomer has layered structure in the similitude of PTFE, it is easy to produce plastic deformation. Thus, it can be inferred that both of them have same wear process on the whole, namely

(1) Debris occurred and plastic accumulation formed on the surface layer of material under the action of microcutting of particles;

(2) The plastic accumulation is lengthened or over lapped by the repeated acting of particles;

(3) Under the further action of particles, the lengthened or over lapped location of the plastic accumulation is ruptured at root and peeled off from the surface.

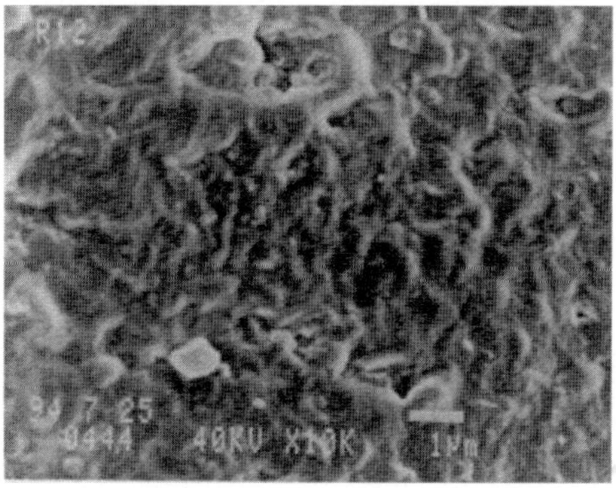

Fig. 6.16 Morphology of the worn surface of FE (in PAM solution)

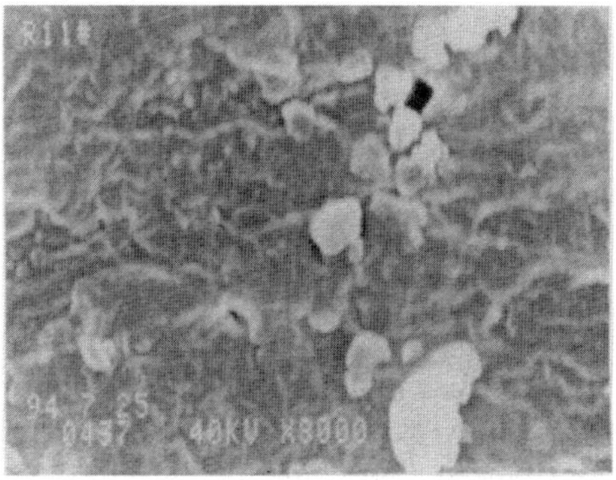

Fig. 6.17 Morphology of the worn surface of FE (in NaOH solution)

The physical model of the above wear processes is illustrated in Figure 6.18.

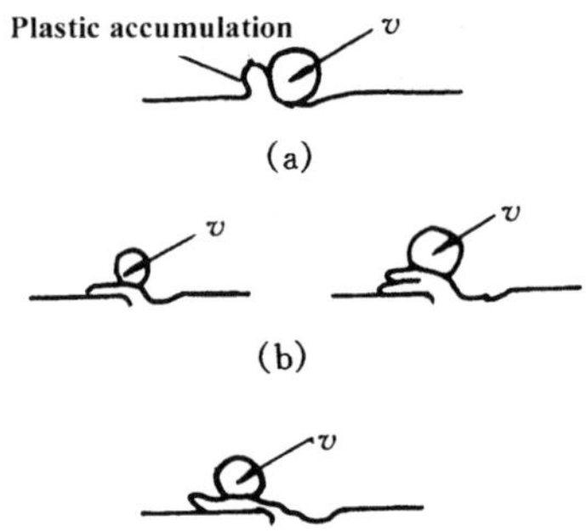

Fig.6.18 Physical model of abrasive erosion of fluoroelastomer
(a)Plastic accumulation (tongue) is produced on the surface layer;
(b)Tongue is lengthened or over lapped;
(c)Tongue is ruptured at root and peeled off from the surface

6.1.3 Wear Speeds

6.1.3.1 Variations of Wear Speeds

On the basis of experimental results, the gravimetric wear speeds of the four kinds of elastomers are given in Table 6.1.

Table 6.1 Gravimetric wear speeds of the four kinds of elastomers (mg/h)

Media	NR	NBR	FE	PU
Water	0.664	3.717	1.021	----
PAM	1.335	6.227	2.269	0.134
NaOH	2.378	8.081	6.021	0.296

As seen from Table 6.1, the wear speed of the NBR material is the most and that of the PU material is the least. It might be accounted for that the tensile strength and tear resistance of the NBR material much more poor than that of the other three kinds of elastomers. Moreover, the nitrile rubber contains polarized groups being hydrolyzed much easy and bi-bonds being oxidized easily. Therefore, its wear resistance is the worst. In addition, it is also found from Table 6.1 that for the rubbers having better elasticity, such as PU and NR, their erosion-resistance is also much better. This is accordance with the results as reached by Hutchings for some unfilled rubbers in erosion experiments [119].

The variations of the above four kinds of elastomers in different media are shown in Figure 6.19.

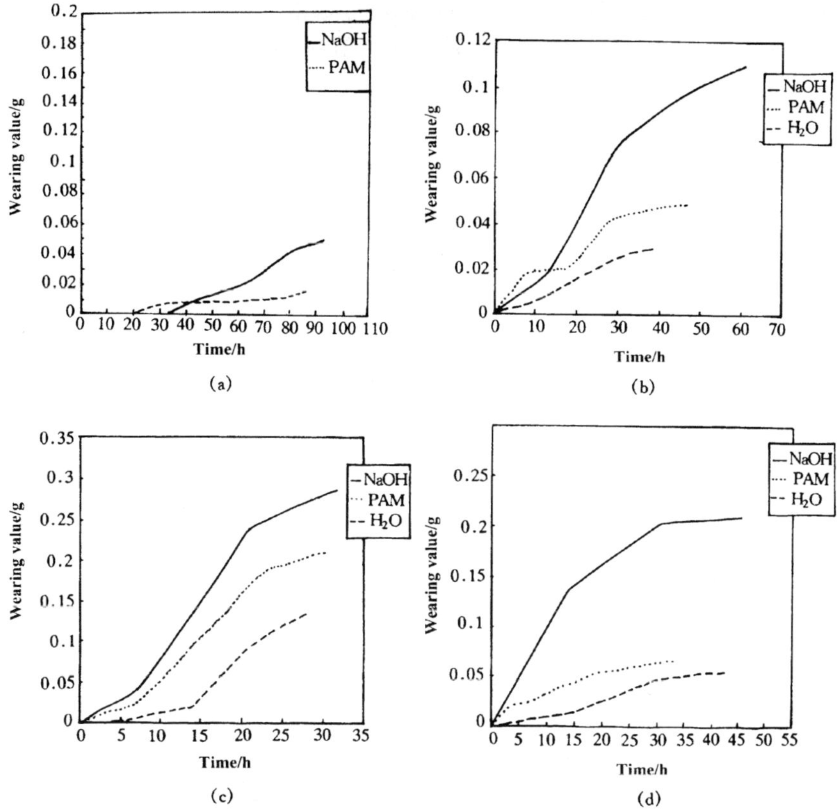

Fig.6.19 Variations of four kinds of elastomers
(a)PU; (b) NR; (c) NBR; (d) FE

As seen from Figure 6.19, the maximum wear speed is found in the medium of NaOH solution for all the four kinds of elastomers. As in this medium, the natural rubber and fluoroelastomer can produced very strong oxidation and the NBR and PU materials are generated much intense hydrolysis. These are the very chemical reactions which cause the molecular chains of rubber to be ruptured more easily, the molecular weight to be reduced obviously and the mechanical property decreasing apparently. As a result, the wearability of elastomers is worse.

The wearing value of elastomers in PAM solution is larger than that in water, the reason is probably that this solution has flocculence to the particles, namely, being in contact with the particles, it can produce hydrogen bonds and electrostatic effect, thus adheres to the surface of particles and causes the particles to gather into a ball. Thus the follow-up performance of the particles is much better. Therefore, the mechanical force acted on the rubber surface by particles is larger and the contact period is also longer, which causes the wearing values to increase.

6.1.3.2 Influence factor [116]

Taking the PU and SBR materials for examples, we will discuss further the effects of flow velocity v, particle size δ and particle concentration C on the wear speeds.

(1) Effect of flow velocity

As seen from Fig.6.20, the wear speeds increase with the flow velocity. Along with the increase in flow velocity, the kinetic energy of the particles enhances. Consequently, the rates of deformation of the rubbers are raised under the action of the particles. In other words, the modulus of elasticity of the materials is enlarged. It would decrease the ability absorbing the impact energy of the elastomers, with the result that the wear speeds of materials are risen.

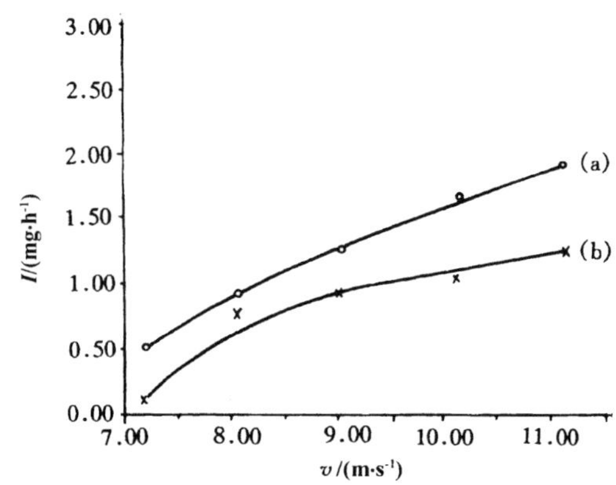

Fig.6.20 Effect of flow velocity on wear speeds

(δ=0.2 ~ 0.3mm, C=10wt%)

(a) SBR (b) PU

In addition, the Reynolds number of liquid flow increase with the velocity of flow, which causes the boundary layer of rubber surface to thin out. Thus, the cushioning effect of the solid surface against the particles decreases. Moreover, a rise in the velocity of flow also increases the frequency acted on the surface of materials by the particles. All of these will raise the wear speeds.

(2) Effect of particle size

As seen from Fig.6.21, the wear speeds increase with the particle size. Obviously, a rise in the particle size is able to enlarge the applied force exerted on the surface of materials by the particles. However, the increase in wear speeds will approach stable after the particle size raises to a certain extent. For the reason that along with the further increase in the size of particles, the velocity of particles reduces and the corresponding kinetic energy decreases. Moreover, the number of particles will be reduced with increasing particle size, provided the particle concentration (particle content of the liquid medium) is unchanged. Thus, the probability of the particles acting on the surface of materials decreases. Therefore, the wear speeds of materials are kept constant as a whole once the particle size rises to a certain extent.

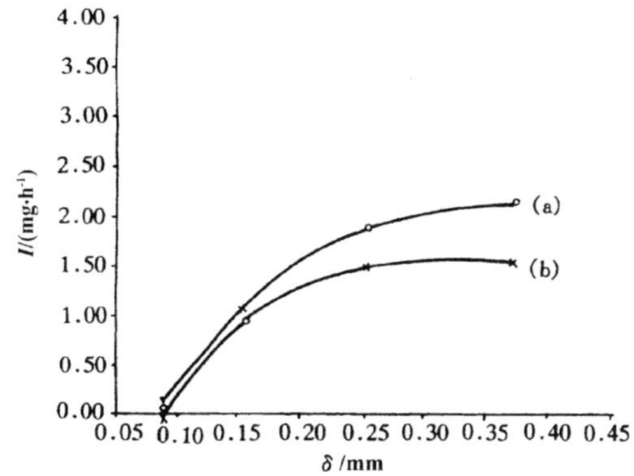

Fig. 6.21 Effect of particle size on wear speeds ($v = 11.08m/s$, $c = 10wt\%$)
(a) SBR (b) PU

(3) Effect of particle concentration

In general, the wear speeds are increased with an increase in the particle concentration, namely the number of particles. However, the denser the particles concentrations, the stronger the interference effect among the particles is, so as to reduce both the probability and the impact angle acted on the surface of materials by the particles. Hence, the wear speeds of materials increase first to a maximum value and then reduce gradually as shown in Fig.6.22.

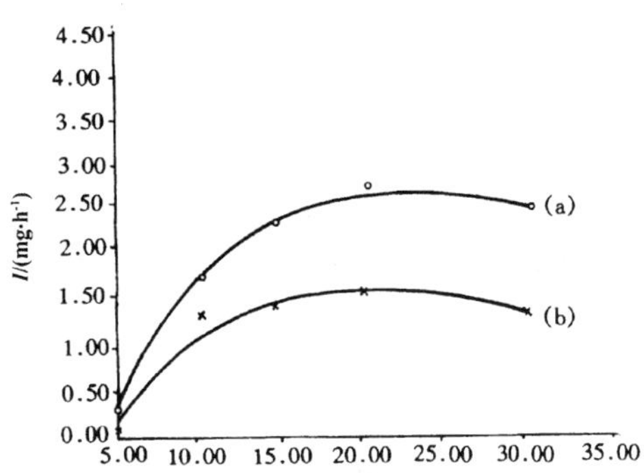

Fig. 6.22 Effect of particle concentration on wear speeds ($v = 11.08m/s, \delta = 0.2 \sim 0.3mm$)
(a) SBR (b) PU

6.1.3.3. Wear equation

Based on the pluralistic linear regression analysis of a great number of experimental data, the wear-speed equation of elastomers being examined can be represented by:

$$I = Kv^{\alpha} C^{\beta} \delta^{\gamma} \tag{6.2}$$

The values of coefficient K, exponents α, β and γ, which depend on the properties of materials being examined, are given in Table 6.2.

Table 6.2 Values of coefficient K, and exponents α , β, γ

Materials	K	α	β	γ
PU	0.4518	2.7985	0.9967	2.0843
SBR	7.728	1.6768	0.8725	2.1219

6.2 A Brief Introduction to Theory of Metal Erosion

Since 1940's, the influence factors on erosion of materials and the corresponding measures increasing the wear resistance of metals were studied extensively. However, the history of research on the theory of erosion is only 30 years. It has been found that the analysis of erosive wear for ductile and brittle materials must apply different theories respectively. In this section, three theories of erosion applied to plastic material are introduced, which could be use to analyze the rubber erosion for reference. Since the variation of eroding value of the elastomers with the impact angle is similar to that of the plastic materials.

6.2.1 Theory of Microcutting

Late in the 1950's, Finnie, Libijef and Blachikov [120~122] respectively proposed their theories of erosion based on the microcutting mechanism of material surface eroded by the abrasive particles almost at the same time. However, the theory of erosion presented by Finnie [121] can describe the correlation between the angle of impact and the eroded rate at shallow angle of impact quantitatively and more perfectly. He considered that the erosion mechanism of ductile material on the condition of low angle of impact is microcutting. The main experimental basis is that the length of the craters occurred on the surface of material is much larger than the depth of the craters after eroding. The craters have length/depth ratios of the order is about 10:1. A model proposed by him is a surface of ductile material cut by a rigid angular grain (Figure 6.23). In this model, the generation of erosion is considered as impacting and striking the surface by the hard particles with mass of m at a certain velocity of v and angle of α, and producing microcutting on the surface, as a result, loss of material occurs. In this theory, assumptions were made as follows:

(1) A geometrically similar configuration of the particles is maintained throughout the period of cutting;

(2) The ratio of the vertical force component on the particle face to the horizontal force component has a constant value in the process of cutting, K=P/F;

(3) The ratio of the depth of contact L to the depth of cut z_t has a constant value during cutting, $\Psi = L / z_t$;

(4) The particle cutting face is of uniform width, which is large compared to the depth of cut;

(5) A constant plastic flow stress σ_y is reached immediately upon impact.

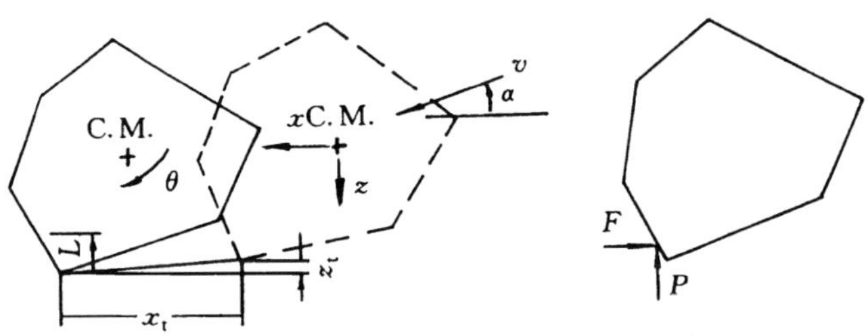

Fig. 6.23 Physical model of microcutting (C.M. is the particles center of gravity) [121]

In addition, it is disregarded that the partial materials are pressed to the both sides of the cut grooves during the process of cutting and the fatigue wear being produced possibly. The expressions for the volume of surface material removed by the abrasive grains of a total mass, M, can be deduced respectively in view of two situations: at lower angles of impact ($\alpha \le \alpha_0 = \arctan\frac{K}{6}$) and at higher angles of impact ($\alpha \ge \alpha_0 = \arctan\frac{K}{6}$). For the former, the particle impacts and cuts out part of the surface, and then the particle tip leaves the surface; as for the latter, the particle tip comes down the surface till its kinetic energy is exhausted. Thus, these expressions are given by

$$V_1 = \frac{Mv^2}{\sigma_Y \varphi K}(\sin 2\alpha - \frac{6}{K}\sin^2\alpha) \quad (\alpha \le \alpha_0) \tag{6.3a}$$

$$V_2 = \frac{Mv^2}{\sigma_Y \varphi K} \cdot (\frac{K}{6}\cos^2\alpha) \quad (\alpha \ge \alpha_0) \tag{6.3b}$$

According to the definition of eroded rate, equation (6.3) can be rewritten as follows:

$$\varepsilon = \frac{\rho v^2}{\sigma_Y \varphi K}(\sin 2\alpha - \frac{6}{K}\sin^2\alpha) \qquad (\alpha \le \alpha_0) \tag{6.4a}$$

$$\varepsilon = \frac{\rho v^2}{\sigma_Y \varphi K} \cdot \frac{K}{6}\cos^2\alpha \qquad (\alpha \ge \alpha_0) \tag{6.4b}$$

The formula of volumetric eroded rate produced by particles per weight was proposed by Libjief [120]. It can be transferred to gravimetric rate of erosion, namely

$$\varepsilon = \frac{\rho f(1-k)}{2g\overline{\sigma}_Y \xi^m} v^2 \sin^2\alpha \cdot (\cot\alpha - fk) \tag{6.5}$$

Where, f —coefficient of friction;

k —coefficient of recovery;

g —gravitational acceleration;

$\overline{\sigma_y}$ —average yield stress of material;

ξ —coefficient of expansion;

m —constant;

The other symbols are ditto.

The theory of microcutting is appropriate to the erosion produced by the multiple angulated particles under the condition of low angle of impact. However, it must much err for the erosion produced by the non-multiple angulated particles (such as round grains) when the angle of impact is larger than 45°.

6.2.2 Theory of Deformation-Cutting [123]

Bitter [123] put forward a theory of deformation-cutting based on the view point of energy balance in the period of erosion. He held that the process of erosion is comprised two parts occurred simultaneously: deformation and cutting action. Therefore, this type of wear must include both deformation wear and cutting wear. At low angle of impact, the cutting wear is dominated, but the deformation wear is dominant at high angle of impact.

The elastic or plastic deformation is probably produced by particles striking the solid surface, which depends on if the impact force causes the solid surface to reach yield limit. On the basis of Hertzian theory, the critical velocity of a ball grain, which impinges perpendicularly on the solid surface and cause the surface to produce plastic deformation, can be given by

$$v_c = \frac{2\sigma_s}{3c} \tag{6.6}$$

Where, σ_s —yield limit of material;

c —constant

Based on the principle of energy equilibrium, the volumetric wearing value of deformation wear can be deduced as

$$V_D = \frac{M}{2u}(v\sin\alpha - K)^2 \tag{6.7}$$

Where, u — deformation wear factor, namely the energy loss per volume of material in deformation wear;

K — constant, i.e. v_c

Similarly, in the view of energy, the volumetric wearing values of cutting wear under two situations are derived as

$$V_{c1} = \frac{2M_c(v\sin\alpha - K)^2}{(v\sin\alpha)^{1/2}}[v\cos\alpha - \frac{c(v\sin\alpha - K)}{(v\sin\alpha)^{1/2}}q] \quad (\alpha \leq \alpha_0) \tag{6.8a}$$

$$V_{c2} = \frac{M}{2q}[v^2\cos^2\alpha - K_1(v\sin\alpha - K)^{3/2}] \quad (\alpha \geq \alpha_0) \tag{6.8b}$$

$$c = \frac{0.286}{\sigma_s}(\frac{\rho_p}{\sigma_s})^{1/4} \tag{6.9}$$

$$K_1 = 0.82\sigma_s^2(\frac{\rho_p}{\sigma_s})^{1/4}(\frac{1-\mu_1^2}{E_1} + \frac{1-\mu_2^2}{E_2})^2 \tag{6.10}$$

Where, q —cutting wear factor; namely the energy loss per volume of material in cutting wear;

c — constant;

K_1 — constant;

ρ_p — density of particle;

μ_1, μ_2 — Poisson ratio;

E_1, E_2 — elastic modulus

Where α_0 is the impact angle while $V_{c_1} = V_{c_2}$, it can be obtained from $\alpha_0 = \arctan\frac{\sigma_s}{2}$. Thus, the total volumetric wearing value is given as follows:

If $\alpha < \alpha_0$

$$V_1 = V_D + V_{c_1} \tag{6.11}$$

If $\alpha > \alpha_0$

$$V_2 = V_D + V_{c_2} \tag{6.12}$$

Based on the definition of eroded rate, and inserting equations (6.7) and (6.8) into the above equations respectively, the expression of eroded rate (gravimetric rate of erosion) can be derived as follows:

$$\varepsilon_1 = \frac{\rho}{2u}(v\sin\alpha - K)^2 + \frac{2\rho(v\sin\alpha - K)^2}{(v\sin\alpha)^{1/2}}[v\cos\alpha - \frac{c(v\sin\alpha - K)}{(v\sin\alpha)^{1/2}}q] \quad (\alpha < \alpha_0) \tag{6.13a}$$

$$\varepsilon_2 = \frac{\rho}{2u}(v\sin\alpha - K)^2 + \frac{\rho}{2q}[v^2\cos^2\alpha - K_1(v\sin\alpha - K)^{3/2}] \quad (\alpha > \alpha_0) \tag{6.13b}$$

This theory of erosion has been proved by some experimental results.

Neilson and Gilchrist [124] presented further simplified equations based on the above theory. The simplified equations can be rewritten as follows:

$$\varepsilon_1 = \frac{1}{2\Psi}(v^2\cos^2\alpha_0 - v_p^2) + \frac{1}{2\varepsilon}(v\sin\alpha_0 - v_{e1})^2 \quad (\alpha \le \alpha_0) \tag{6.14a}$$

$$\varepsilon_2 = \frac{1}{2\Psi}v^2\cos^2\alpha_0 + \frac{1}{2\varepsilon}(v\sin\alpha_0 - v_{e1})^2 \quad (\alpha > \alpha_0) \tag{6.14b}$$

Where, ψ, ε are constants, they represent the critical kinetic energy which must be absorbed by the surface to release one unit mass of eroded material for cutting and deformation wear respectively;

v_p — horizontal component of rebound velocity of particle after the particle impacting the worn surface;

v_{el} — vertical component of velocity, being caused the worn surface of material to produce elastic deformation.

The above equations are more simply described the two kinds of eroded characteristics resulted from microcutting and deformation.

6.2.3 Fatigue Theory [120]

According to this theory, the erosive wear is resulted mainly from the frictional fatigue rather than the action of microcutting, namely, the fatigue damage of the frictional surface is ascribed to the action of both normal and tangential forces during frictional process. This fatigue is different from the bulk or contact fatigue of materials. However, this fatigue damage is similar to the usual fatigue of material when the elastic interaction is occurred between the particles and the surface of materials. If there exists plastic interaction instead of elastic one, the fatigue is close to low-cycle fatigue. Based on this fatigue theory, the equations of the erosive wear-rate by weight under elastic and plastic contact conditions can be obtained respectively.

For elastic contact, we have

$$\varepsilon_e = K_{t_1} \rho^{t/5} \rho_p (\cos\alpha - f)(v \sin\alpha)^{2+\frac{2t}{5}} (\frac{kf}{\sigma_0})^t (\frac{4}{3\pi\Theta})^{\frac{4t}{5}-1} \tag{6.15}$$

The above equation is applicable to rubber and plastics.

For plastic contact, we have

$$\varepsilon_p = K_{t_2} \rho^{\frac{t}{5}} (\frac{\rho}{3})^{\frac{t+1}{4}} (\frac{v \sin\alpha}{\sqrt{c\sigma_s}})^{\frac{t+5}{2}} [\frac{2}{e}\sqrt{\frac{1+kf}{1-kf}}]^t (\cos\alpha - f) \tag{6.16}$$

Where, f — frictional coefficient of contact area;

Θ — elastic modulus of material, $\Theta = (1-\mu)/E$;

e — relative deformation;

k — proportional coefficient, $k = 2H_{HB}/\sigma_s$;

c — coefficient being considered the change of effective yield-limit resulted probably from the cold hardening and proportional coefficient;

t — characteristic index of fatigue, for the plastic contact conditions, generally, t =2;

σ_0 — frictional fatigue strength, whose value is obtained by extrapolating the frictional fatigue curve to the value n =1; On certain condition, it is close to the limiting value of the strength of materials. n is the number of cycles leading to removal of wear particles;

E — elastic modulus of material;

μ — Poisson ratio of material;

H_{HB}— hardness of material;

σ_s— yield-limit of material;

K_{t_1}, K_{t_2} — coefficients which are the function of Poisson ratio depends on the applied strength theory.

$$K_{t_1} = \sqrt{\pi}\frac{t}{2t+5}\frac{\Gamma(\frac{t}{5})}{\Gamma(\frac{t}{5}+0.5)} \tag{6.17}$$

$$K_{t_2} = \sqrt{\frac{\pi}{2}}\frac{\Gamma(\frac{t+5}{4})}{\Gamma(\frac{t+7}{4})} \tag{6.18}$$

Where Γ is Gamma function.

The other symbols are the same as mentioned previously.

It should be pointed out that the contact stress and deformation are increased with the increase in the impact angle α, and if once they reach the damaging values, the wear mechanism of material must be changed from frictional fatigue to microcutting. The critical impact angle α_{kp} leading to this changing can be estimated according to the following equations:

Elastic contact

$$\alpha_{kp} = \sin^{-1}[(\frac{\sigma_0}{4kf})^{\frac{5}{2}} \frac{\pi^2\Theta^2}{v\sqrt{5\rho}}] \tag{6.19}$$

Plastic contact

$$\alpha_{kp} = \sin^{-1}[\frac{3e^2}{4v}(\frac{1-kf}{1+kf}\sqrt{\frac{3c\sigma_s}{\rho}})] \tag{6.20}$$

Apparently, equations (6.15) and (6.16) can be applicable only if $\alpha < \alpha_{kp}$.

The three erosion theories as stated above might be applied to analyze the effects of the properties of material and operating conditions on erosion, yet these theories are mainly considered the wear process of material resulted from the erosion by solid particles, which is probably a most common process.

6.3 Theory of Abrasive Erosion

6.3.1 Theoretical Equation of Eroded Wear-Rate [125]

In consideration of that the main difference between the abrasive erosion and the impact erosion is the magnitude distribution of impact angle of the particles, thus the theory of abrasive erosion might be set up on the basis of the above theories of metal erosion.

The three kinds of theoretical equations of wear rate deduced by the theories of microcutting, deformation-cutting and fatigue respectively as mentioned above can be unitedly expressed by a general equation as follows:

$$\varepsilon = Qf(\alpha) \tag{6.21}$$

Where, α is impact angle and the factor Q is dependent on the erosion theory to be concerned.

It is assumed that the magnitude of impact angles is distributed according to the distribution function $\Phi(\alpha)$ during abrasive eroding process. $\Phi(\alpha)$ is a density function of impact angle α, $0^\circ < \alpha_1 < \alpha < \alpha_2 < 90^\circ$.

Thus, a general equation of wear-rate of abrasive erosion can be given by:

$$\varepsilon_\alpha = \int_{\alpha_1}^{\alpha_2} \varepsilon\Phi(\alpha)d\alpha \tag{6.22}$$

Based on the theory of fluid mechanics, the pulsating velocity v_i induced by the turbulent flow is conformed to the normal distribution, its density function is as follows:

$$\varphi(v_i)=\frac{1}{\sqrt{2\pi}}\exp[-\frac{1}{2}(\frac{v_i^2}{\sigma})] \qquad -\infty < v_i < +\infty \tag{6.23}$$

Where, σ is the variance of v_i, namely $\sigma=(\overline{v_i^2})^{0.5}$, it is related to the Reynolds number, viscous friction, velocity of flow, region of turbulent flow being occurred and so on.

In the flow field examined (Figure 6.24), v, v_r, v_i are the tangential and radial velocities as well as the radial-pulsating velocity of the fluid units respectively. If we neglect the influence of the circumferential and axial pulsating velocities of fluid on the impact angle and presume that the particles have good following character, namely, moving in step with the fluid and no relative motion between them, we can obtain

$$\tan\alpha=(v_r+v_i)/v \tag{6.24}$$

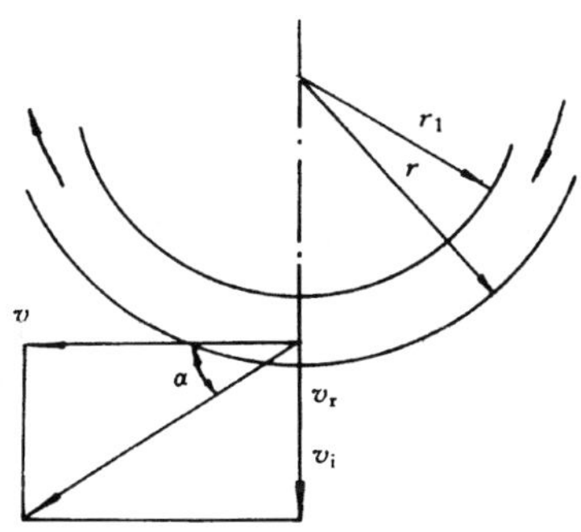

Fig. 6.24 Velocity of fluid unit

If the rotating fluid is considered as a planar circulation flow, then $v_r/v=c$ and c is a constant less than unity. Thus, the above equation can be rewritten as

$$\tan\alpha=c+(v_i/v) \tag{6.25}$$

Where, $c+(v_i/v)>0$, namely, while $+\infty > v_i > -vc$, $\alpha>0$.

Based on the probability theory, if the distribution of v_i is known, the density function of α can be given as

$$\Phi(\alpha)=\frac{v}{\sqrt{2\pi}}\sigma\cos^2\alpha\exp[-v^2(\tan\alpha-c)^2/2\sigma^2] \qquad 0^\circ\le\alpha\le 90^\circ \tag{6.26}$$

By putting the above equation into equation (4.22), we obtain

$$\varepsilon_a = \int_{\alpha_1}^{\alpha_2} \varepsilon(\frac{v}{\sqrt{2\pi}}\sigma\cos^2\alpha)\exp[\frac{-v^2(\tan\alpha - c)^2}{2\sigma^2}]d\alpha \tag{6.27}$$

Substituting for the theoretical equations of wear rate from the theories of erosion as described in the Section 6.2, i.e., equations (6.4), (6.5), (6.13), (6.15) and (6.16) into the above equation, the corresponding theoretical equations of wear rate of abrasive erosion will be obtained as follows.

(1) Based on microcutting theory and from equations (6.4) and (6.5), we have

$$\varepsilon_a = \int_{\alpha_1}^{\alpha_0} \frac{\rho v^2}{\sigma_Y \psi K}(\sin 2\alpha - \frac{6}{K}\sin^2\alpha)\frac{v}{\sqrt{2\pi}}\sigma\cos^2\alpha\exp[-v^2(\tan\alpha - c)^2/2\sigma^2]d\alpha +$$
$$\int_{\alpha_0}^{\alpha_2} \frac{\rho v^2}{\sigma_Y \psi K}(\frac{K}{6}\cos^2\alpha)\frac{v}{\sqrt{2\pi}}\sigma\cos^2\alpha\exp[-v^2(\tan\alpha - c)^2/2\sigma^2]d\alpha \tag{6.28}$$

and

$$\varepsilon_a = \int_{\alpha_1}^{\alpha_0} \frac{\rho f(1-k)}{2g\sigma_Y \xi^m} v^2\sin^2\alpha(\cot\alpha - fk)\frac{v}{\sqrt{2\pi}}\sigma\cos^2\alpha\exp[-v^2(\tan\alpha - c)^2/2\sigma^2]d\alpha \tag{6.29}$$

(2) Based on deformation-cutting theory and from equation (6.13), we have

$$\varepsilon_a = \int_{\alpha_1}^{\alpha_0} \frac{\rho}{2u}(v\sin\alpha - K)^2 + \frac{2\rho(v\sin\alpha - K)^2}{(v\sin\alpha)^{1/2}}[v\cos\alpha - \frac{c(v\sin\alpha - K)}{(v\sin\alpha)^{1/2}}q]\times$$
$$\frac{v}{\sqrt{2\pi}}\sigma\cos^2\alpha\exp[-v^2(\tan\alpha - c)^2/2\sigma^2]d\alpha +$$
$$\int_{\alpha_0}^{\frac{\pi}{2}} \frac{\rho}{2u}(v\sin\alpha - K)^2 + \frac{\rho}{2q}[v^2\cos^2\alpha - K_1(v\sin\alpha - K)^{3/2}]\times$$
$$\frac{v}{\sqrt{2\pi}}\sigma\cos^2\alpha\exp[-v^2(\tan\alpha - c)^2/2\sigma^2]d\alpha \tag{6.30}$$

(3) Based on fatigue theory and from the elastic contact equation (6.15), we obtain

$$\varepsilon_{ae} = \int_{\alpha_1}^{\alpha_2} K_{t_1}\rho^{\frac{t}{5}}\rho_p(\cot\alpha - f)(v\sin\alpha)^{2+\frac{2}{5}t}(\frac{kf}{\sigma_0})^t \times$$
$$(\frac{4}{3\pi\Theta})^{\frac{4}{5}t-1}(\frac{v}{\sqrt{2\pi}}\sigma\cos^2\alpha\exp[-v^2(\tan\alpha - c)^2/2\sigma^2]d\alpha \tag{6.31}$$

From the plastic contact equation (6.16), we have

$$\varepsilon_{ap} = \int_{\alpha_1}^{\alpha_2} K_{t_2}\rho_p(\frac{\rho}{3})^{\frac{t+1}{4}}(\frac{v\sin\alpha}{\sqrt{c\sigma_s}})^{\frac{t+5}{2}}\left[\frac{2}{e}\sqrt{\frac{1+kf}{1-kf}}\right]^t(\cot\alpha - f)\times$$
$$\frac{v}{\sqrt{2\pi}}\sigma\cos^2\alpha\exp[-v^2(\tan\alpha - c)^2/2\sigma^2]d\alpha \tag{6.32}$$

On the basis of fatigue theory an expression to determine the wear speed of rubber was presented by Stalel and Latnel [126]:

$$I = \rho^{\frac{1}{5b+1}} (v \sin\alpha)^{\frac{2}{5b+2}} (\frac{E}{1-\mu^2})^{\frac{4}{5b-1}} (\frac{Kf}{\sigma})^b (cot\alpha - f) \quad (6.33)$$

Where, b — coefficient of kinetic fatigue;

μ — Poisson ratio;

σ — tearing strength;

K — constant.

The other symbols are the same as the above.

The above equation was proved by the laboratory tests of some materials, in which the tearing strength, coefficient of kinetic fatigue and frictional coefficient are introduced. The coefficient of friction represents indirectly the elasticity of rubber as it is determined by the hysteresis of rubber basically.

6.3.2 An Energy Approach to the Prediction of Erosive Wear-Rate [127]

This is an approach to determine the erosive wear of equipment and pipeline transporting two-phase (liquid-solid) flow. Its fundamental is outlined as follows:

(1) The amount of material removed is assumed proportional to the mechanical energy dissipated by the particle-wall interaction, the coefficient of proportionality is dependent on the wear pattern and the presence of other surface processes, such as corrosion, cavitation, scaling and other erosion mechanisms.

(2) The worn material caused by erosion is assumed to be removed by one of the following three mechanisms: directional impact, random collisions and Coulombic friction (Figure 6.25). As for the predominant type of wear mechanism is determined by the two-phase flow pattern.

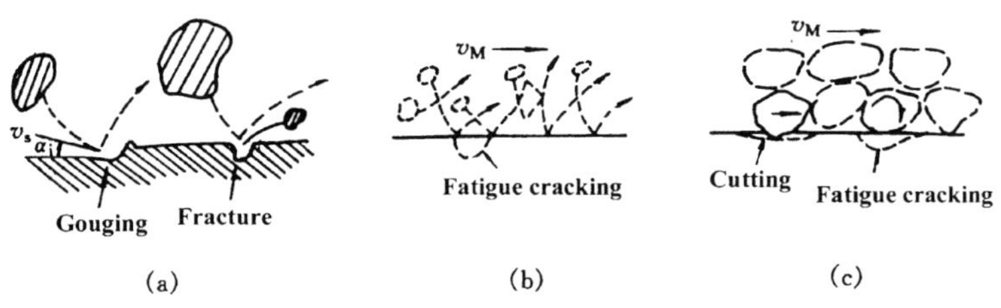

Fig.6.25 Wear mechanisms caused by particle-wall interaction [127]
(a) Directional impact; (b) Random collisions; (c) Coulombic friction

(3) The erosion mechanisms have a stochastic character: particles of random size and shape slide, roll and impact on to the wall with random velocities under various impact angles, while the non-homogeneities of the exposed material have a random distribution. Therefore, the material removal is characterized by a probability function.

(4) Experimental and computational steps are combined to determine the eroded wear-rate, namely, small-scale experiments in the laboratory provide the empirical correlations between the wear rate and energy dissipated for the above three wear patterns for any pair of materials (solid particles/worn wall) [128], and then, the concentration and velocity distributions in the areas of interest, and particularly close to the exposed walls, are determined by using numerical calculation methods in fluid dynamics [129, 130].

The total wear rate, namely loss in wall thickness per unit time, averaged in time, is estimated by using the correlations between the amount of material removed and interaction energy for each of the above three wear mechanisms. It can be expressed as follows:

$$\Delta \dot{s} = \sum_{i=v,k,fr} \Delta \dot{s}_i = \sum_{i=v,k,fr} \phi'_i (\dot{E}_i - \dot{E}_{i0}) \tag{6.34}$$

Where, v, k, fr — corresponding to the three wear mechanisms: directional impact, random impact caused by the kinetic energy and frictional wear;

ϕ'_i — coefficient showing the differential effect of the ith wear mechanisms, for a given pair of materials (particles/worn wall) when all other surface processes are invariant;

$\dot{E}_i$ — time rate of energy dissipation by the ith wear mechanism;

$\dot{E}_{i0}$ — threshold energy rate for incipient wear

The computational part of the algorithm is shown in the flowchart "WEAR" (Fig.6.26).

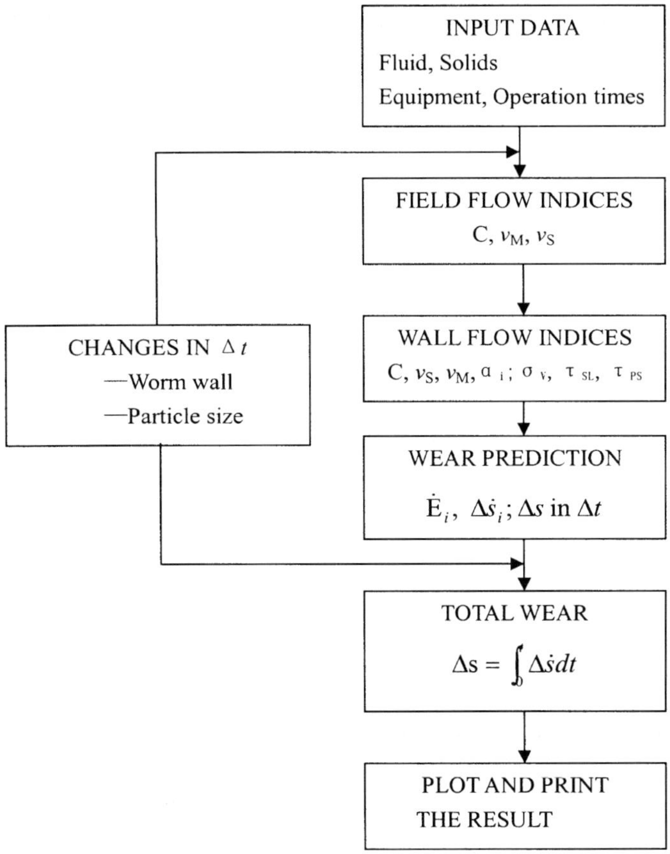

Fig.6.26 Flowchart "WEAR" [127]

The "INPUT DATA" includes the fluid properties and solid characteristics; the dimensions, material and operation indices of the equipment; and the operation time, The main "FIELD FLOW INDICES" computed in the code are: solid concentration C, mixture velocity v_M and solid velocity v_S. From this, the "WALL FLOW INDICES" of significance for the wear process is obtained, including the wall distributions for concentration *C*, impact solid velocity v_S and impact angle α_i, as well as the solid stresses caused by Coulombic friction τ_{SL} (supported load stress), random impingement due to the velocity fluctuations τ_{DS} (dispersive stress) and directional impact σ_v (dynamic pressure caused by solid particles) [131]. With these indices the "WEAR PREDICTION" can be made.

The total wear rate is

$$\dot{\Delta s} = \sum \dot{\Delta s_i} = \dot{\Delta s_v} + \dot{\Delta s_k} + \dot{\Delta s_{fr}} \tag{6.35}$$

and

$$\dot{\Delta s_v} = \Phi_v(\alpha_i)[v_S(\sigma_v - \sigma_{v0})] \tag{6.36}$$

$$\dot{\Delta s_k} = \Phi_k[v_{S,tg}(\tau_{DS} - \tau_{DS_0})] \tag{6.37}$$

$$\dot{\Delta s_{fr}} = \Phi_{fr}[v_{S,tg}(\tau_{SL} - \tau_{SL_0})] \tag{6.38}$$

Where, $\dot{\Delta s_v}$ — wear rate caused by the velocity fluctuations;

$\dot{\Delta s_k}$ — wear rate caused by the fluctuation kinetic energy of particles (random velocity fluctuations);

$\dot{\Delta s_{fr}}$ — wear rate caused by the sliding and rolling friction of particles;

$\Phi_v(\alpha_i)$ — loss in thickness per unit energy dissipated per unit area by directional impact under impact angle α_i;

Φ_k — loss in thickness per unit energy dissipated per unit area by random impact of solids;

Φ_{fr} — loss in thickness per unit energy dissipated per unit area by sliding and rolling friction;

$v_{S,tg}$ — tangential component of the solid velocity to the worn wall;

$\sigma_{v_0}, \tau_{DS_0}, \tau_{SL_0}$ — threshold values for incipient wear

$$\sigma_v = C\rho_p v_S^2/2 \tag{6.39}$$

Here ρ_p is the density of particle.

The mixture velocity v_M can be computed by using turbulence models, or inviscid flow simulations with slip boundary conditions when the simplification is acceptable [128].

The wear in Δt_j alters the wall geometry, wall roughness and indirectly the flow indices. Therefore, an iterative loop in the computer code ("CHANCES IN Δt") is required (Figure 6.26).

However, the probabilistic approach can not lead alone to a full answer to the wear prediction, in part due to the complexity and the limitation of the actual knowledge on the wear processes at the microscopic level. Thus, an experimental approach is adopted, namely, the typical wear patterns are individually simulated in the laboratory, from where the specific losses in wall thickness per unit of removed material $\Phi_v, \Phi_k, \Phi_{fr}$ are determined.

This energy approach incorporates the mechanistic and probabilistic analyses to describe the particle-wall interactions at particle scale. It can be applied to determine the wear rate in equipment and pipeline handling slurries for different wear patterns, various material characteristics and flow configurations. It can be also used to determine the location of maximum wear and the operating life of the equipment. This approach for the wear rate has been applied to straight pipes of variable slope and to centrifugal pump casings [127].

6.3.3 A Fracture Mechanics Approach to the Prediction of Erosive Wear-Rate

Recently, Arnold and Hutchings investigated the erosive wear of unfilled elastomers [132], the fatigue crack growth is considered as the main wear mechanism. Moreover, they have been found that the wear mechanism of erosion at low impact-angle being termed glancing impact by them is very similar in nature to that of the abrasive wear by a sharp blade or by a smooth indenter. A series of ridges, running transversely to the impact direction, is produced during the initial stages of erosion. Therefore, similar to the abrasion by a blade sliding against to the rubber surface, Arnold and Hutchings [133] presented a physical model of the material removal by a hard particle operating at glancing impact of the fine particles based on the simplified theory of abrasion proposed by Southern and Thomas [54] as described in Chapter 5. This model is shown in Figure 6.27.

Several assumptions were made as follows while the mathematical model was developed:

(1) All the frictional force is carried by one ridge at a time;

(2) The impacting particles are spherical in shape;

(3) Large deformations of the surface occur but embedment of particles does not;

(4) The motion of the particles is pure sliding throughout the contact period;

(5) The coefficient of friction is constant;

(6) Viscoelastic effects can be neglected.

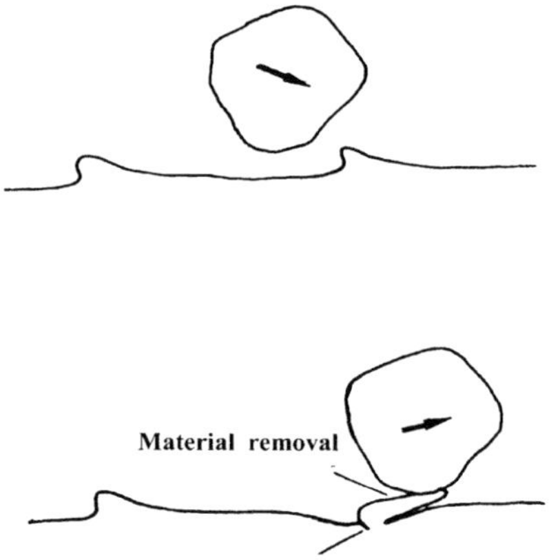

Fig. 6.27 A physical model of impact of a particle on a ridge [133]

Based on the equation (5.1), a theoretical expression of eroded wear-rate can be derived as follows:

$$\varepsilon = 2(\frac{8}{3\pi})^{(\beta-1)/2} B \sin\theta(1+\cos\theta)^{\beta} \rho_p^{(\beta-1)/2} v_0^{\beta+1} R^{\beta-1} f^{\beta} \rho_r (\frac{E}{1-\mu^2})^{(\beta-1)/2} Q \quad (6.40)$$

Where, β — constant relating crack-growth character of rubber, that is the α in equation (5.1);

v_0 — initial velocity of the impacting particle;

α — impact angle;

R— radius of the particle;

f — coefficient of sliding friction;

ρ_p — density of the particle;

ρ_r — density of the rubber;

E — Young's modulus of the rubber;

μ — Poisson ratio of the rubber;

Q— function of the impact angle;

The other symbols are all the same to those in equation (5.1).

While the impact angle is smaller, corresponding to abrasive erosion, or the frictional coefficient is less, such that for $f \tan\alpha < 0.5$, we have

$$Q = \sin\alpha^{\beta}(\cos\alpha - f\sin\alpha) \quad (6.41)$$

While the impact angle is larger, corresponding to impact erosion, or the frictional coefficient is higher, such that for $f \tan\alpha > 0.5$, we have

$$Q = \frac{\pi^{\beta-1}}{2^{\beta}} \sin^{\beta} \alpha[\cos\alpha - f \sin\alpha(1 - \frac{\sin kt_c}{kt_c})]kt_c(\frac{1-\cos kt_c}{kt_c})^{\beta} \quad (6.42)$$

Where kt_c is given by

$$kt_c = \arccos(1 - \frac{1}{f \tan\alpha}) \quad (6.43)$$

and

$$k = [2ER / m(1-\mu^2)]^{1/2} \quad (6.44)$$

Here t_c is the contact time, at which the tangential velocity of particles becomes zero. It must be less than the time of contact between the particles and the solid surface under the condition of erosion. m is the mass of a particle.

As seen from the equation (6.40), the erosion rate of rubber is mainly dependent on the property of rubber, particle size as well as the impact velocity and impact angle.

Therefore, the expression presented for the erosive wear rate can be used to predict the erosive wear behavior of elastomers qualitatively. The erosion rate is highest at a low angle, moreover, the value of this angle increases with the value of β. The erosion rate is also raised with the increase in impact velocity and particle size. However, lubrication causes a reduction in the erosion rate, as does a lower elastic modulus.

This approach is unable to predict the erosive wear rate quantitatively, probably due to several assumptions implicit in the theoretical expression. Obviously, certain assumptions are unreasonable, such as, assuming that the frictional coefficient and elastic modulus of rubber are unchangeable, neglecting the influence of particle size and particle flowing and so forth on the erosive wear rate.

6.3.4 Theoretical Expression for Prediction of Erosive Wear-Rate in Annular Pipes [134]

6.3.4.1 Kinetic Analysis of Particles in Annular Flow Field

A calculating model presented (Figure 6.28) is used as the basis of the following kinetic analysis of particles in annular flow field.

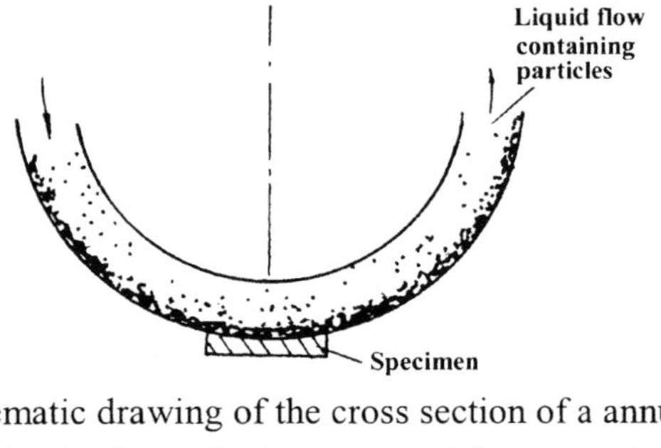

Fig. 6.28 Schematic drawing of the cross section of a annular pipe

Generally, the particles in flow field are exerted by a number of forces, among which only the larger is taken into account, namely, the fluid drag applied on the particles. The other smaller forces, including the applied force of the flow-field pressure gradient exerted on the particles, floatage, gravity, etc. are all neglected. Moreover, it is assumed that all the particles are spheroid with same size. The mutual impinging among the particles during flow process is out of consideration. Thus, the particle trajectories in flow field can be determined according to the theory of fluid mechanics.

In consideration of the effect of low Reynolds number on adjacent position of the pipe wall, the viscous resistance is larger and can not be ignored. Moreover, the kinetic analysis of the particles involves the velocity analysis of flow field. Hence, the flow field should be divided into two parts, the inside and the outside of the boundary layer, to be calculated respectively.

1) Boundary layer

According to Newton's second law, the kinetic equation for particles exerted by the fluid drag is given by

$$m_p du / dt = 3\pi\eta\delta(v - u) \tag{6.45}$$

Then

$$du / dt = 3\pi\eta\delta(v - u) / m_p \tag{6.46}$$

Where, m_p — mass of particle;

t — time;

u — velocity vector of particle;

v — velocity vector of fluid;

δ — particle size;

η — viscosity of fluid;

The velocity distribution of flow field is expressed as a logarithmic curve.

2) Outside of boundary layer

The kinetic equation for particles is

$$m_p \frac{du}{dt} = \frac{C_D A_s \rho_f (v - u)|v - u|}{2} \tag{6.47}$$

Then

$$\frac{du}{dt} = \frac{C_D A_s \rho_f (v - u)|v - u|}{2m_p} \tag{6.48}$$

Where, C_D — viscosity factor;

A_x — projective area being vertical to the direction of fluid flow;

ρ_f — density of fluid.

The velocity of flow field is linear distribution based on the measured values of the flow field in annular pipe.

Applying cylindrical coordinate (r,θ,z) and taking the z axis coincided with the central line of the annular pipe, on condition that the influence of inlet of the flowing path is out of consideration, the motion of fluid is stable according to the symmetrization in the directions of both θ and z axes, thus

$$\begin{cases} v_r = 0 \\ v_z = 0 \\ v_\theta = v_\theta(r) \end{cases} \tag{6.49}$$

Where, v_r, v_z, v_θ are the components of fluid velocity in the three directions of cylindrical coordinate respectively. The velocity distribution of the outside of boundary layer is given by

$$v_\theta(r) = r(2\bar{v})/(r_1 + r_2) \tag{6.50}$$

Where, r — radius of flow field;

r_1 and r_2 — internal and outer radii of the flowing path respectively;

$\bar{v}$ — average velocity of the fluid in flowing path.

Rewriting equations (6.46) and (6.48) as follows:

$$du/dt = G(v-u) \tag{6.51}$$

For boundary layer

$$G = \frac{3\pi\delta\eta}{2m_p} \tag{6.52}$$

For the outside of boundary layer

$$G = \frac{C_D A_s \rho_f |v-u|}{2m_p} \tag{6.53}$$

Using cylindrical coordinate (θ,r,z), gives

$$\begin{cases} u_\theta = rd\theta/dt \\ u_r = dr/dt \\ u_z = dz/dt \end{cases} \tag{6.54}$$

Then equation (6.51) can be rewritten as follows:

$$\begin{cases} \dfrac{du_\theta}{dt} + \dfrac{u_r u_\theta}{r} = G(v_\theta - u_\theta) \\ \dfrac{du_r}{dt} - \dfrac{u_\theta^2}{r} = G(v_r - u_r) \\ \dfrac{du_z}{dt} = G(v_z - u_z) \end{cases} \tag{6.55}$$

Assuming that only u_θ, u_r and u_z are varied with time in a minimal time Δt, the following equations can be obtained by integrating equations (6.54) and (6.55) respectively:

$$\begin{cases} \theta = \theta_0 + u_\theta \Delta t / r \\ r = r_0 + u_r \Delta t \\ z = z_0 + u_z \Delta t \end{cases} \tag{6.56}$$

and

$$\begin{cases} u_\theta = v_\theta - \dfrac{u_{r0} v_{\theta 0}}{G r_0 + u_{r0}} + (u_{\theta 0} - v_\theta + \dfrac{u_{r0} v_{\theta 0}}{G r_0 + u_{r0}}) \exp[-(G + u_{r0}/r_0)\Delta t] \\ u_r = v_r + \dfrac{u_{r0}^2}{G r_0} + (u_{r0} - v_r + \dfrac{u_{\theta 0}^2}{G r_0}) \exp[-G\Delta t] \\ u_z = v_z + (u_z - v_z)\exp[-G\Delta t] \end{cases} \tag{6.57}$$

Where, u_θ, u_r and u_z are the components of velocity of particle in the three directions of cylindrical coordinate respectively.

Inserting equation (6.57) into equation (6.56), the moving trajectories in flow field of particles can be determined.

The numerical calculation of the moving trajectories of particle in flow field for a cone of hydrocyclone with 203.2mm in diameter being applied to drilling in oil field under the operating conditions corresponding to the severe-wear site is shown in Figure 6.29 and its partial data are listed in Table 6.3.

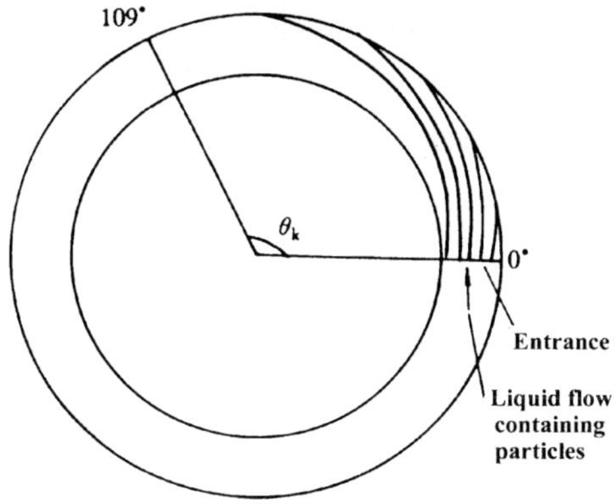

Fig. 6.29 Moving trajectories of particles in flow field

Table 6.3 Velocity and position of particle impacting on pipe wall

Tangential velocity (m/s)	Radical velocity (m/s)	Impacting position (degree)	Number of impacting
6.442	0.1550	96.24	1
2.876	0.0576	104.17	2
2.241	0.0249	107.24	3
2.055	0.0113	108.59	4
1.986	0.0057	109.23	5
1.955	0.0033	109.57	6
1.942	0.0015	109.73	7

As shown from Fig.6.29 and Table 6.3, under the action of centrifugal force, the particle deviates from the direction of fluid flowing and impacts on the pipe wall along the direction of the outer radius of fluid path. While the particle bounding back, its radical velocity drops quickly due to the action of fluid drag and centrifugal force exerted on the particle in the radical direction. After repeating this process for several times, the radical velocity of the particle approaches to zero gradually, and then the particle will slide and roll closely against the pipe wall. It can be also seen from the Table 6.3 that the velocity of the particle impacting the pipe wall for the first time is much higher than that occurred later on.

That angle where the particle being located initially at the internal radius of flowing path begins to slide and roll along with the pipe wall may be termed dividing angle or critical angle θ_k (Fig. 6.29). It means that beyond this angle, the motion of the particle is only sliding and rolling along with the pipe wall and without impacting. Therefore, the flowing process of the two-phase flow in the annular pipe can be divided in two stages according to the moving mode of the particle being relative to the pipe wall. In the first stage (within the scope of θ_k), the relative motion of the particle against the pipe wall is mainly impact. In the second stage (beyond the scope of θ_k), the relative motion in question is mainly sliding or accompanied partial rolling.

The prime factors of influence on the dividing angle are discussed below:

(1) Particle size.

The effect of particle size on angle θ_k is shown in Figure 6.30.

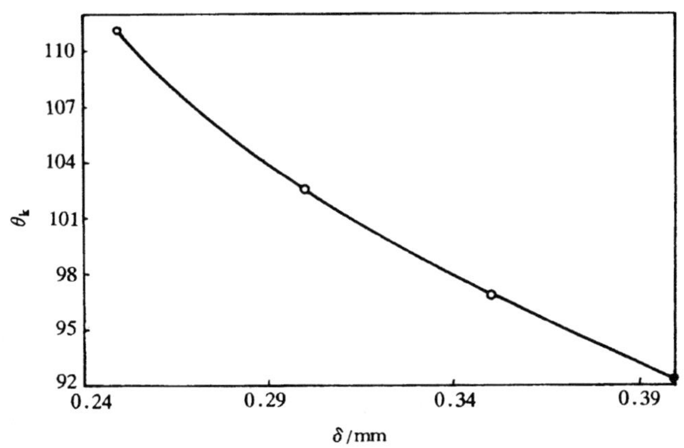

Fig.6.30 Variation of angle θ_k with particle size

As seen from Fig.6.30, the larger the particle size δ, the smaller the angle θ_k is, which is almost changed linearly. The reason is that the fluid drag and the centrifugal force exerted on the particles are proportional to the second and third power of δ respectively; therefore, the larger the particle size is, under the action of centrifugal force the particle will overcome the fluid drag more easily, and will slide being accompanied by rolling along with the pipe wall more early. As a result, the angle θ_k is smaller.

(2) Relative density of particles.

The effect of particles' relative density on angle θ_k is shown in Figure 6.31.

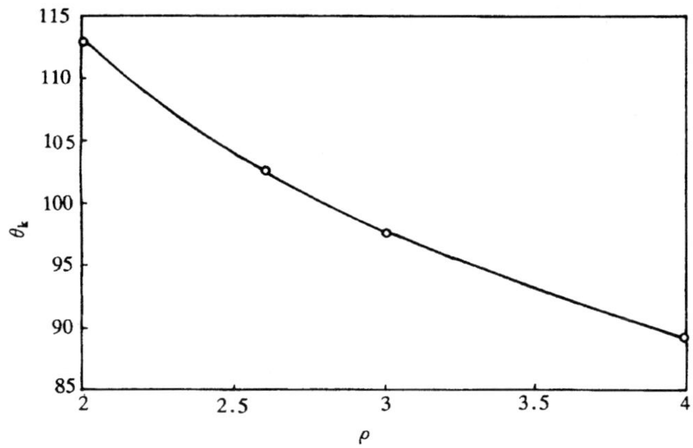

Fig. 6.31 Variation of angle θ_k with relative density of particle

The relative density of particles ρ is the ratio of particle density ρ_p to liquid density ρ_f, namely $\rho=\rho_p/\rho_f$. As seen from Fig.6.31, the larger the relative density of particles, the smaller the angle θ_k is. This correlation is almost linear. Moreover, this effect is accounted for that the centrifugal force exerted on particle is proportional to ρ_p and is nothing to the fluid drag applied on particle. Therefore, the larger the ρ is, the particle will overcome the fluid drag and slide being accompanied by rolling against the wall of pipe more easily, which causes the angle θ_k to decrease.

(3) Fluid velocity.

As seen from Fig.6.32, the larger the fluid velocity v, the smaller the angle θ_k is. Because the higher the fluid velocity, the larger the fluid drag exerted on the particle is, that can raise the tangential speed of the particle and increase the centrifugal force applied on the particle. The combination effect of the centrifugal force and the fluid drag on the particle results in the particle sliding and rolling against the pipe wall more easily, namely angle θ_k decreases. However, the extent of decrease in θ_k is more gently in low-velocity portion than that in high-velocity portion.

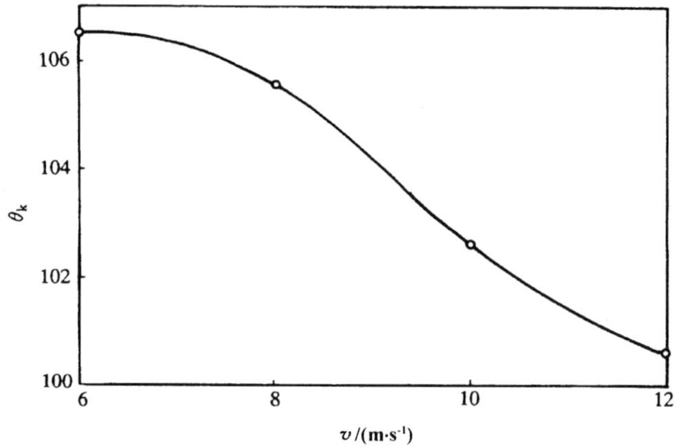

Fig. 6.32 Variation of angle θ_k with fluid velocity

(4) Geometry of flowing path.

The geometry of flowing path is characterized by the width-to-radius ratio R, namely the ratio between width D and outer radius r_2 of the flowing path. As seen from Table 6.4, the angle θ_k is risen with the increase in R. Obviously, the larger R is, the wider the flowing path will be. The particle must have a longer distance to go to the internal wall of pipe at the outer radius of flowing path. So the angle θ_k increases.

Table 6.4 Effect of geometry of the flowing path on angle θ_k

R	1/8	1/4	3/8
θ_k (°)	63.95	102.60	147.95

6.3.4.2 Theoretical Equation of Wear Rate of Abrasive Erosion

(1) First stage

Based on the above analysis, in the first stage of abrasive erosion at the internal wall of annular pipe, though the impact angle of fluid against the pipe wall is zero, the follow-up performance of particle is bad because there has a big difference of density between the particle and fluid. Therefore, the effect of particle on the pipe wall is mainly impacting. Thus the primary type of wear produced in this stage might be considered as impact erosion.

In consideration of that polyurethane has higher hardness and its compressive and tearing strengths are much larger than the contact stress of particle on the pipe wall, then the contact between the particle and pipe wall is elastic contact. Therefore, the theoretical equation of wear rate can be derived using fatigue theory for elastic contact.

Let $t = 3$ [135], this equation can be given as follows from equation (6.15):

$$\varepsilon_e = K_{t_1} \rho^{3/5} \rho_p (\cot\alpha - f)(v \sin\alpha)^{3.2} (\frac{kf}{\sigma_0})^3 (\frac{4}{3\pi\theta})^{1.4} \quad (6.58)$$

(2) Second stage

The effects of particles on the inner wall of pipe are mainly sliding friction and partial rolling friction resulted from the sliding and rolling of particles against the pipe wall during the second stage of abrasive erosion in annular pipe. In the meantime, the stress and deformation occur in the surface layer of pipe wall under the action of normal and tangential forces.

For the same site of the pipe wall, it is acted by different particles back and forth, and then is exerted the cycle stresses repeatedly. Therefore, after a number of stress cycles, the fatigue crack will initiate at the weakest site of the surface or subsurface layer. Then, the crack propagates and wear debris appears at last. Moreover, the liquid will accelerate the crack growth. Therefore, the wear mechanism of abrasive erosion in annular pipe during the second stage is fatigue delamination resulted from friction of particles in fluid flow against pipe wall, which has been proved by the experimental results of abrasive erosion of PU materials [116].

Based on the wear mechanism as stated above, in order to derive the theoretical equation of wear rate, the criterion of fatigue damage, namely the number of stress cycle resulted in fatigue wear of materials n, should be determined at first [136]:

$$n = (\sigma_0 / \sigma)^t \quad (6.59)$$

Where, σ_0— stress to damage for material being undergone one stress cycle, in general, it is the limit of tensile strength of materials;

σ— cyclic tensile stress;

t— exponent of fatigue characteristic.

The cyclic tensile stress is given by [114]:

$$\sigma = 2f\sigma_{max} \quad (6.60)$$

and

$$\sigma_{max} = 1.5\bar{\sigma} \quad (6.61)$$

Where, f— sliding frictional coefficient;

$\bar{\sigma}$— average contact stress;

σ_{max}— maximum contact stress on the surface of pipe wall.

Substituting equations (6.60) and (6.61) into equation (6.59), the n can be obtained by

$$n = (\frac{\sigma_0}{3f\bar{\sigma}})^t \quad (6.62)$$

The volumetric wearing speed per unit time can be expressed as follows:

$$I = V_p \frac{N}{n} K \tag{6.63}$$

Where, V_p — volume of a debris;

N — number of particles sliding over on the inner wall of pipe per unit time;

K — probability of particle sliding and rolling on the pipe wall.

Generally, the debris is in cylindrical form. According to the necessary condition to produce debris, namely the elastic deformation-energy equals to the surface energy [137]. Thus, we have

$$V_p = \frac{2\pi a^2 \gamma E}{\bar{\sigma}^2} \tag{6.64}$$

Where, a — radius of contact region of particle against the pipe wall;

γ — surface energy per unit area of material of pipe wall;

E — modulus of elasticity of debris.

Assuming that the particles sliding and rolling on the inner wall of pipe are distributed over the corresponding pipe wall fully and evenly, thus the probability K of particle sliding and rolling on the pipe wall can be obtained by

$$K = \frac{8Cm_p}{\pi\delta^2 S_c C_w \rho_f} \tag{6.65}$$

$$C = hR_2 / (R_1 + R_2) \tag{6.66}$$

Where, S_c — cross-section area of flow passage;

C_w — weight concentration of particles;

C — characteristic coefficient of flow passage;

h — height of flow passage;

R_1, R_2 — inner and outer radii of annular pipe.

The other symbols are just the same as stated above.

As for the number of particles sliding over on the inner wall of pipe per unit time, it is given by

$$N = \frac{\bar{v} S_c C_w \rho_f}{m_p} \tag{6.67}$$

Where, $\bar{v}$ is the average velocity of fluid in flow passage.

Inserting equations (6.64), (6.65) and (6.67) into equation (6.63), the theoretical equation for wearing speed is obtained as follows:

$$I = \frac{16C\gamma E\bar{v}}{\bar{\sigma}^2} \left(\frac{a}{\delta}\right)^2 \left(\frac{3f\bar{\sigma}}{\sigma_0}\right)^t \tag{6.68}$$

The cycle stress σ exerting on the pipe wall, i.e., average contact stress being applied on the pipe wall during the particles sliding and rolling along with the wall under the action of fluid flow is given by

$$\sigma = \frac{1}{4} p_f (\frac{\delta}{a})^2 \tag{6.69}$$

Where, p_f is the pressure exerted by the centrifugal force of fluid flow on pipe wall, namely

$$p_f = 2\rho_f \bar{v}^2 (\frac{R_2 - R_1}{R_2 + R_1}) \tag{6.70}$$

As a result, we have

$$\sigma = \frac{1}{2} \rho_f \bar{v}^2 \frac{R_2 - R_1}{R_2 + R_1} (\frac{\delta}{a})^2 \tag{6.71}$$

For PU materials, taking t=3 and substituting equation (6.71) into equation (6.68), the theoretical equation of volumetric wearing speed can be derived finally as follows:

$$I = 216C' E\gamma\rho_f f^3 \bar{v}^3 / \sigma_0^3 \tag{6.72}$$

Where

$$C' = C\frac{R_2 - R_1}{R_2 + R_1} \tag{6.73}$$

Equation (6.73) indicates the relationship among the wearing speed and the physical-mechanical properties of material, operating conditions as well as the frictional performance. The calculating results based on this equation are in accord with the experimental data obtained in References [116]. The equation above can be used to predict the wearing speed of PU materials.

As for the theoretical calculation of wearing speed of abrasive erosion in the inner wall of straight pipes under the condition of pressure transmission of two phases of solid-liquid fluid, it could be considered as a special case of the calculation of wearing speed of abrasive erosion in annular pipes. The wear process is still fatigue wear resulted from friction.

6.3.4. Application of the Energy Theory of Rubber Abrasion to Analyzing the Mechanism of Abrasive Erosion of Rubber

According to the energy theory of rubber abrasion, as stated in Section 5.2 of Chapter 5, it is thought that partial elastic energy of the rubber surface might be converted into the rupture energy in the process of a single particle sliding on the rubber surface during abrasive erosion.

On the basis of equation (6.67), the expression to calculate the number N of particles sliding over the inner wall per unit time can be rewritten as

$$N = \frac{\bar{v}C_w}{\delta^3} \tag{6.74}$$

Where, $\bar{v}$ - average velocity of fluid flow;

C_w - weight concentration of fluid;

δ - diameter of particle.

The probability of particle sliding on the inner wall is given by [134]:

$$K' \propto \frac{\delta}{C_w} \tag{6.75}$$

Where K' is the probability of particle sliding on the inner wall.

Then, the stored potential energy of the material per unit time is expressed as

$$U = K'NE_f \tag{6.76}$$

Where E_f is ruptured energy.

The wear energy density e* can be obtained as follows by use of the criterion volume of fatigue damage, i.e., equation (6.59).

$$e^* = \frac{nE_f}{V_p} \tag{6.77}$$

Where, n is the number of stress cycles. It can cause a certain volume of material V_p to form wear debris. This volume of wear-debris V_p is able to calculate by equation (6.64).

Based on the concept of wear energy density expressed as equation (5.34), we have,

$$V = U / e^* \tag{6.78}$$

Inserting equations (6.76) and (6.77) into the above equation, it is obtained that

$$V = K'NV_p / n \tag{6.79}$$

Then substituting equations (6.74), (6.75), (6.61) and (6.64) into the above equation, and taking t=3 for PU materials [134], we have

$$V \propto \bar{v}^{-3} r / \sigma_0^3 \tag{6.80}$$

As will be readily seen from the above equation, the wearing volume is proportional to an exponent of three of the average velocity of fluid flow, and is inversely proportional to an exponent of negative three of the tensile strength. It is in accord with the conclusion drawn in the above section, namely equation (6.72) perfectly.

Chapter 7

FATIGUE WEAR AND FRICTIONAL WEAR

Besides the abrasive wear and erosive wear discussed above, the basic types of rubber wear are also included fatigue wear and frictional wear.

Fatigue wear is such a kind of wear in the same way as the abrasion, which is produced against a rough counterface, but the difference between them is that the rough counterface for the former has blunt projections and that for the latter has sharp texture.

As for the frictional wear, it is a specific type of wear for rubber, which takes place on the smooth counterfaces having a high coefficient of friction.

7.1 Fatigue Wear

7.1.1 Wear Mechanism

As early as the late fifties in the last century, Kragelskii [138] presented the concept of fatigue wear. He definitely pointed out that fatigue is one of the sources to cause wear. Fatigue wear of rubber is a relatively widespread type of wear generated between the rubber and its counterface with hard and blunt asperities on condition that the friction force and contacting stress are not too large. It is a kind of low-intensity wear by comparison with the abrasive wear. Since this wear is much similar to the abrasive wear in many respects, it is regarded as a type of abrasive wear by some scholars. For example, Schallamach called it abrasion on blunt abrasives [84].

Since the rubber is a highly elastic material, when it moves relatively to a rigid and rough surface with blunt rather than sharp asperities under a certain normal load, it will support cyclic loading from beginning to end and produce repeated compression, expansion and reversed shear stresses within the surface layers. As a result, the surface layers of rubber will wear through. Therefore, the basic character of fatigue wear of rubber is that the surface layer of material is damaged under the action of compressive, tensile and shear deformations for many times. These deformations are caused by the interaction of rubber with the hard and blunt projections on the rough surface during sliding.

It should be pointed out that the above fatigue process is limited within the thin surface layer and it becomes complicated owing to the influence of medium of environment. It is worn through rather slow, but it cannot be neglected under the condition of long- term existence of cyclical load and rather small adhesive action. This wear generally displays slight abrasive wear and tearing of surface layer of material, but no wear pattern appears on the worn surface.

Fatigue wear of rubber occurs as a result of repeated deformations. Under the given abrasion conditions, the average number of reversed dynamic stress cycles (namely the number of deformation) n required for the surface layer of rubber to be damaged and separated from the substrate is a function of the fatigue resistance of rubber and the stress state determined by the normal load, sliding velocity, geometry of friction surface and other factors. It is a kind of characteristic parameters to reflect the fatigue resistance of rubber. It is supposed that the following empirical equation is valid [13]:

$$n = (\sigma_0/\sigma)^t \tag{7-1}$$

Where, $_0$ - tensile strength of the elastomer in a simple tensile test;

σ - amplitude of the repeated dynamic or cyclic tensile stress;

t - characteristic index of dynamic fatigue or coefficient of fatigue resistance depending on the character and surface behavior of material.

Because index t has practically nothing to do with temperature (50-120℃), concentration of stress(sample with quantitative cutting edge) and frequency (below 50HZ)[139], it is much conveniently applied to characterize the fatigue property of rubber. The characteristic index of dynamic fatigue of several kinds of rubber is given in Table 7.1.

Table 7.1 Evaluation of characteristic index of dynamic fatigue for various rubbers [140]

Materials	SBR(30-70)	SBR(at 5℃)	NR	Sodium-Polybutadiene
t	1.9	2.2	1.75	1.4

7.1.2 Rate of wear

Several assumptions are made as follows:

(1) A number of idealized sinusoidal asperities are evenly distributed on a rigid rough counterface;

(2) The wear volume of rubber is proportional to its deformed volume;

(3) The thickness of the worn layer is in proportion to the depth of penetration of asperities and some measure of the surface roughness effect;

(4) The amplitude of the repeated dynamic or cyclic stress is proportional to the mean normal load on an asperity.

Based on the above presumptions, the following equation can be obtained [13]:

$$\frac{\beta}{f} = K\sigma_o^t E^{2(1-t)/3}\left(\frac{N}{A}\right)^{(1-t)/3}\left(\frac{\lambda}{R}\right)^{(5-2t)/3} \tag{7-2}$$

Where, β - coefficient of abrasion resistance;

f - coefficient of friction;

K - constant;

E – elastic modulus;

N - normal load;

A - actual contacting area;

λ- mean wavelength of the surface asperities;

R – radius of the asperity tips.

Equation (7-2) reflects the fatigue-wear resistance of rubber and characterizes the approximately quantitative relation among the behaviors of frictional couple and the basic test parameters. It might be used to calculate the coefficient of abrasion resistance β. Thus it can be seen that the wear resistance of rubber depends on the tensile strength σ_0, module of elasticity E and the characteristic index of dynamic fatigue t of the rubber as well as the frictional coefficient f of the frictional couple. In addition, the normal load N and the effects of both velocity and strength introduced indirectly by changing the variety of rubber have also influence on the wear resistance of rubber.

Despite some hypotheses are made while deriving the above equation, it is very useful in clarifying the nature of fatigue wear, and it has been confirmed experimentally in certain cases. Moreover, every item of equation (7-2) has definite physical meaning, which can be measured experimentally. Thus this equation is able to apply to define the reasonable operating conditions of frictional couple and choose the material of rubber with best comprehensive mechanical properties. For instance, for a given set of properties (i.e., a given rubber), this equation is simplified as follows:

$$\frac{\beta}{f} = CN^{(1-t)/3} \tag{7-3}$$

Where, C is constant.

The above equation shows that for a given elastomer, β/f depends on the normal load alone, namely it decreases with the increase in normal load. This equation has been verified experimentally [13].

In order to more accurately explain the relation among the roughness of frictional surface, the wear speed I and the mechanical properties of rubber, an expression was given as follows [141]:

$$I = k\left(\frac{fE}{\sigma_0}\right)^t \left(\frac{N}{E}\right)^{1+t\gamma} \tag{7-4}$$

Where, k - constant;

γ - parameter of roughness, being equal to $1/(2\nu+1)$;ν is an exponent in equation of curvature of bearing surface. For tarred and concrete roads, ν=3.0, γ =0.14.

From equations (7-2) and (7-4), it can be seen that the fatigue wear is increased with the increase in elastic modulus of the rubber and the normal load, and with the decrease in the tensile strength and the characteristic index of dynamic fatigue t.

7.2 Frictional Wear

When rubber with a relatively low tearing strength slides on smooth counterfaces having a relatively high coefficient of friction a new mechanism of wear[87] was discovered, which causes roll formation at the sliding interface and eventual tearing of the rolled fragment. This type of wear is called frictional wear and also termed wear caused by roll formation, or rolling wear by some scientists [13, 19, 87, 142].

7.2.1 Wear mechanism

When a stiff body with smooth surface contacts with a projection on the rubber block under the action of a normal load N and moves relatively and parallelly to the rubber surface with a horizontal velocity v (Figure 7.1, (a)), the corresponding physical process of occurrence of frictional wear of rubber can be described as follows [13,87]:

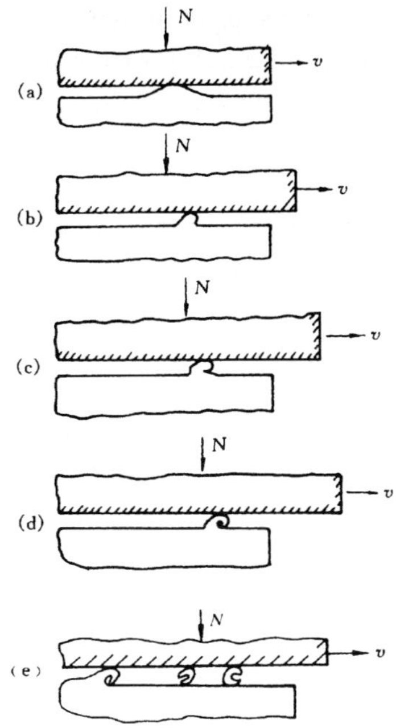

Fig.7.1 Physical process of frictional wear of rubber

(1) If the tangential force is not large enough to overcome the frictional force at the interface, no slipping of the rubber on the contact surface occurs but the projection of rubber will be deformed, and this deformation will be severe more and more with the increasing in friction at the interface (Figure 7.1, (b)).

(2) If once the tangential force at the interface exceeds the frictional resistance, the body will move relatively on the rubber surface. However, if the frictional force is very large and the tear strength of rubber is relatively low, a part of the surface layer of rubber with severe deformation could be torn before slippage at the interface takes place when this part of layer is in a state of maximum strain, and a cut or crack may appear perpendicular to the direction of attempted sliding (Fig.7.1, (c)). This cut is usually initiated at that spot of the rubber surface where the largest stretching action is subjected. However, if once the crack produces, it will grow gradually under the action of a less stress. The local direction of the crack growth depends in much complex fashion on the nature of the local stress condition and a series of other factors, such as the molecular heterogeneities in the structure of the rubber and the unevenness of temperature distribution at the interface nearby, etc. It is unlikely that the subsequent growth of these cuts will result in immediate separation of wear debris from the surface layer of rubber. A much more likely event is the gradual tearing of the rubber, so that a little relative movement in the contact zone is possible without complete slippage.

(3) The tearing part of the surface layer of rubber is separated off further to form a strip of tongue and to wind into a roll (Fig.7.1, (d)), which is assumed to be in a stressed condition. The force, which causes elongation due to adhesion, depends on the tear resistance of the rubber at the place where the shred separates off from the surface layer of rubber. Failure of a shred occurs when the elongation reaches a critical maximum value, and the result is detachment of the rolled shred in the form of a roll from the surface layer. The elongation of the shred depends on its cross-sectioned dimensions, which are generally variable, and a series of the factors determining the direction of crack growth.

(4) After the rolled shred is formed, the stiff body is put in relative motion under the conditions of rolling friction caused by the rolled debris (Fig.7.1, (e)), in the mean time, the surface layer of rubber is continuous to be tore, rolled, stretched and separated off, thereby some free rolled debris of the rubber are accumulated between the stiff body and the surface of rubber.

Obviously, frictional wear can take place only when there exists a certain combination of external conditions and the properties of the rubber being abraded. This type of wear is more likely for the rubber with a low tear strength and the high coefficient of friction existing at the interface. Tear strength of the rubber depends mainly on the temperature resulted from sliding friction, and in certain cases frictional heating may cause the surface layer of rubber resinification and tackiness, which adds to the frictional coefficient.

In accordance with the above analysis, the total frictional power P_f used to elongate and tear off shred from the surface layer of rubber can be given by [13]:

$$P_f = P_t + P_e + P_H \tag{7-5}$$

Where, P_f - power used to tear shred of rubber from the surface layer;

P_e - power used to elongate shred of the rubber;

P_H - power used in hysteresis losses which accompany roll formation.

It is clear that frictional wear can take place only when the power losses of sliding friction at interface is larger than P_f, so the main condition which determines the probable occurrence of this type of wear may be expressed as follows [13]:

$$P_f \leq fNv \tag{7-6}$$

Where, f is the coefficient of static friction at the interface.

As seen from above, a main feature of the mechanism of frictional wear is that the frictional work used to form the wear debris by frictional wear is far larger than that by any other kinds of wear.

Although the mechanism of frictional wear as stated above is by no means either complete or fully authenticated. However, as seen from equation (7-6), it is very likely that under certain conditions it predominates. The plot of gravimetric wear rate vs. load for natural rubber sliding on abrasive paper and on hard rubber (shore hardness 84) is shown in Fig.7.2 respectively, which shows the contributions of various mechanisms of abrasion to the overall wear rate under different conditions. In the case of the abrasive paper, wear is due primarily to fatigue wear at low loads, with the contribution of abrasive wear becoming more dominant according as the normal load is increased. As for the natural rubber-on-hard rubber combination, the mechanism of wear is probably due to frictional wear. It is seen that at low loads the abrasive paper contributes substantially more wear than the natural rubber-hard rubber combination. When the normal load increases to a certain value, the wear rate of natural rubber-on-hard rubber combination is far larger than that of natural rubber-on abrasive paper combination [13]. Experiments have shown [87] that when the coefficient of friction exceeds the value of about 1.15, there is a rapid and drastic increase in wear rate. This phenomenon is satisfactorily explained by the mechanism of frictional wear and just verified that the equation (7-6) is correct. It means that the frictional wear predominates. In practice, it is certain that a combination of all three types of wear, namely fatigue wear, frictional wear and abrasive wear, usually take place simultaneously, and it is very difficult to isolate the separate contributions of each form of wear to the overall wear effect[13].

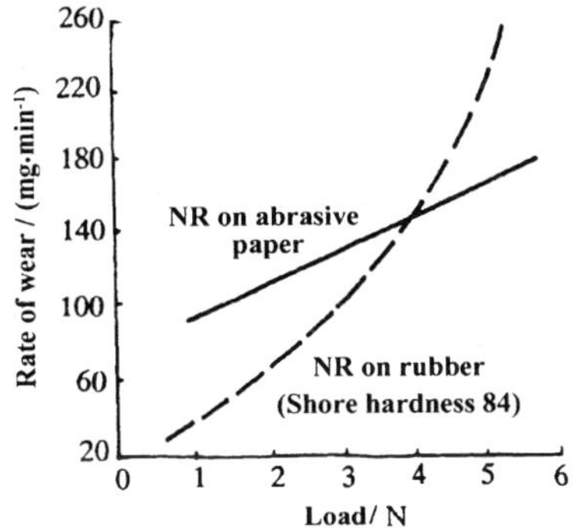

Fig. 7.2 Load dependence of wear rate for natural rubber [13, 19]

7.2.2 Rate of wear

In accordance with the equations (3-7), (3-8), (3-10) in *Chapter 3* and equation (7-6), the coefficient of abrasion resistance β of the rubber can be rewritten as following expression:

$$\beta = P_f / I \tag{7-7}$$

Where, P_f - frictional power;

I - wear speed.

On the basis of the concept of volumetric wear speed, namely the equation (3-10) in *Chapter 3*, β can be also rewritten on a time basis as follows [13]:

$$\beta = P_f / (d\Delta V / dt) \tag{7-8}$$

Where, (dV/dt) is the volumetric wear per unit time, namely volumetric wear speed.

The components P_t, P_e and P_H in equation (7-5) can be expressed in terms of existing theory [87], so that eventually equation (7-8) has the general form as follows [13]:

$$\beta = (T, E, R_d, \delta, b, r) \tag{7-9}$$

Where, T - characteristic tear energy of the rubber;

E - elastic modulus of the rubber;

R_d - dynamitic resilience of the rubber (it equals to the ratio of energy returned to a system during a half-cycle of vibration to the energy expended);

δ, b, r - the average values of the thickness, width and radius of a shredded roll respectively.

In addition, based on the concept of characteristic energy for tearing [143,144] and a series of studies in the loss of work of the rubber being rolled on the surface of stiff body [142], the coefficient of abrasion resistance can be given as follows:

$$\beta = \frac{T}{\delta} + \omega + 0.9 \times \frac{N^{4/3}(1-϶)(\varepsilon+1)}{\delta b(Erb)^{1/3}} \tag{7-10}$$

Where, ω- average value of unit tensile energy;

϶ - rubber elasticity;

ε- percentage elongation.

The other symbols are the same with that in equation (7-9).

Though some hypotheses are made in derivation of the above equation, this equation can clarify the relation among the frictional wear resistance and the elasticity-relaxation property as well as the strength of the rubber. Based upon this relation, the critical condition of frictional wear being generated can be defined.

Chapter 8

SURFACIAL MECHANOCHEMICAL EFFECTS OF ABRASIVE EROSION

8.1 Introduction

8.1.1 Research Objects of Mechanochemistry and Tribochemistry

At the beginning of last century the term of mechanochemistry was introduced by Ostwald [145] to define a new scientific field concerned with the effect of mechanical energy on chemical reactions. After several years, the area of dispersion or deflocculation by mechanical means rather than by chemical but involving the use of principles in physical chemistry was called "mechanochemistry" by Travis [146]. However, only in the last four decades, mechanochemistry has developed gradually into an independent branch of science. It is a peripheral subject being shaped up on the basis of intersection of mechanics and chemistry.

Heinicke [147] defined the mechanochemistry as "a branch of chemistry dealing with the chemical and physical-chemical changes of substances of all states of aggregation due to the influence of mechanical energy". Essentially, the mechanochemistry is a science involved with the processes of mechanical energy transferring to chemical energy. Therefore, we hold that mechanochemistry is a science dealt with various chemical and physicochemical reactions initiated of all kinds of substances being excited state by the mechanical action, such as impacting, squeezing, pulling and friction, etc. By the excited state, it means a state of thermodynamical non-equilibrium with higher energy and reactional activity than those of thermodynamical equilibrium in the view of thermodynamics. However, in a broad sense, it is considered by some scientists as a science to investigate the processes of mutual transforming between the internal mechanical energy and chemical energy of substantial system [148].

As seen from the developing process of this science, the term "mechanochemistry" was initially used to describe the effect of mechanical action on chemical reactions, namely mechanochemical effects. Later on, a new effect of chemical reactions on the mechanical performance of the solid and some mechanical phenomena, which could be referred to as "chemomechanical" effects was discovered. In this connection, Gutman [149] held that to avoid introducing another new term 'chemomechanics', it is necessary to understand the term 'mechanochemistry' in both its meanings, including all types of mutual transformations of mechanical and chemical energy.

In 1960's, Boramboim [150] established a new branch of chemistry, namely mechochemistry of polymers, which is a peripheral discipline involved with both of mechanics and polymer chemistry. It deals with mechanochemical reactions of polymers including various chemical and physicochemical changes resulted from mechanical actions, such as crack, structurization, cyclization, ionization and isomerzation etc.

Late in the 1960's Heinide [147] coined a term "tribochemistry" and advanced its definition: "Tribochemistry is a branch of chemistry dealing with the chemical and physico-chemical changes of solids due to the influence of mechanical energy." However, some scientists considered that the mechanochemistry is precisely the tribochemistry in tribology field [44] or might stand for the tribochemistry [151]. Strictly speaking, tribochemistry deals with various chemical and physicochemical reactions being induced by the excited solid surfaces in relative motion under the action of mechanical forces. It mainly involves the chemical and physicochemical reactions produced between the frictional surfaces of solid, and between these solid surfaces and the media during the process of friction, such as, lubrication of frictional couples and the above reactions in the process of cold-working or hot-working of metal. Therefore, the tribochemistry is a peripheral subject of tribology and chemistry. It is a branch of both chemistry and tribology. Obviously, tribochemistry must not be equated with mechanochemistry, although it is placed in the research realm of mechanochemistry, which is just as mechanics, optical, thermodynamics and so on are included in the study scope of physics.

Kajdas and co-workers [6] have given a more concrete definition of tribochemistry: "It is a branch of chemistry which deals with the chemical reactions in the friction zone. The reactions cause mechanical and physicochemical changes of the surface layer of mating pairs. The reactions involved are caused by different types of energy and catalysis. The most important are chemical interactions of lubricant components with mating surfaces of rubbing elements. The lubricants include oils, greases, solid materials and gases.... Generally, it can be said that tribochemistry deals with the relations between tribomechanics and chemical changes in the elements of the tribological system."

8.1.2 Surfacial Mechanochemistry and Surfacial Mechanochemical Effects

Surfacial mechanochemistry deals with a variety of chemical reactions and physicochemical changes generated in the two-phase interface of various substances which are in excited state owing to the mechanical action. It mainly involves the area of mechanochemistry of solids connected with surface phenomena. Therefore, it could be considered as a branch of mechanochemistry, which is shaped up based on the intersection of surfacial science and mechanochemistry. The research realm of surfacial mechanochemistry intervene mechanochemistry and tribochemistry.

The surfacial mechanochemical effect (reaction) implies varieties of mechanochemical reactions occurring on the two-phase (including a solid phase at least) interfaces. As for the tribochemical reaction, it is the chemical reactions occurring between mating elements and the environment during friction, which resulted in the formation of new products, as described by Kajdas et al. [6].

8.1.3 Surfacial Mechanochemical Effects of Abrasive Erosion

The surfacial mechanochemical effects of rubber occurred in the process of abrasive erosion have mainly four kinds as follows:

(1) Rupture of macromolecular chains;

(2) Surfacial oxygenated degradation;

(3) Hydrolysis;

(4) Thermal decomposition.

In this section, some recent achievements reached by the present author and his group are introduced, which involve the surfacial mechanochemical effects of nature rubber (NR), nitrile rubber (NBR), fluororubber and polyurethane eroded respectively in three different fluid media containing quartz particles, namely, water (H_2O, pH7), polyacrylamid soluteion (PAM, pH6) and sodium hydroxide solution (NaOH, pH12). The related experiments were conducted by using a special abrasive erosion testing machine, as described in Chapter 6, Section 6.1.1.

8.2 Surfacial Mechanochemical Effects of Nature Rubber

Nature rubber (NR) is a macromolecular compound consisted of chain units of isoprene. Its structure of molecular chains is given by:

$$\left[CH_2-\underset{\displaystyle CH_3}{\overset{|}{C}}=CH-CH_2 \right]_n \qquad (n\approx 10000)$$

As seen, the main component of NR is rubber hydrocarbon. The elasticity, tearing strength and wear resistance of NR are much better, but its oxidation resistance and ozonation resistance are worse. Moreover, this kind of rubber is easily aged.

By the analysis of binding energy and spectra of the original and worn surfaces of NR material, it could be inferred that there are two surfacial mechanochemical effects on the worn surfaces eroded in the three different media respectively.

8.2.1 Fracture of Macromolecular Chains

The decreasement of methlylene groups (-CH_2) and methyl groups (-CH_3) (Fig. 8.1) shows that the fracture of macromolecular chains occurs and the products with low molecular weight and low disparity as well as the small molecular and monomers are produced; moreover, the new crosslinked structures are formed.

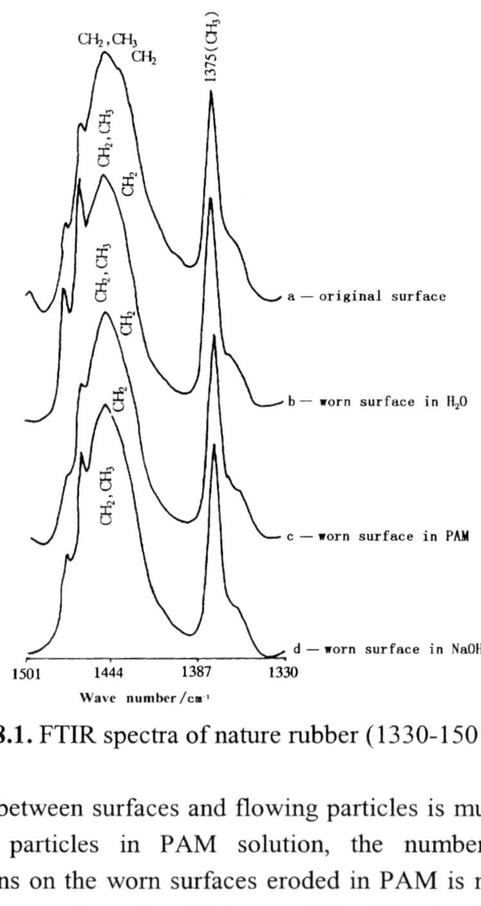

Fig.8.1. FTIR spectra of nature rubber (1330-1501cm^{-1})

The interaction between surfaces and flowing particles is much stronger due to the cohesive action of particles in PAM solution, the number of the fracture of macromolecular chains on the worn surfaces eroded in PAM is much more. Moreover, the fractures of quite a number of multiple crosslinked bonds of sulphur are produced besides the carbon-carbon bond cleavage. These conclusions can be proved by the XPS and FTIR analyses. It has been identified that the sulphur oxygen covalent bonds (-S-O-) existed (Fig.8.2) and the degree of decreasement of CH_2 groups is much more on the worn surface relative to the original surface (Fig.8.3 and Fig.8.4).

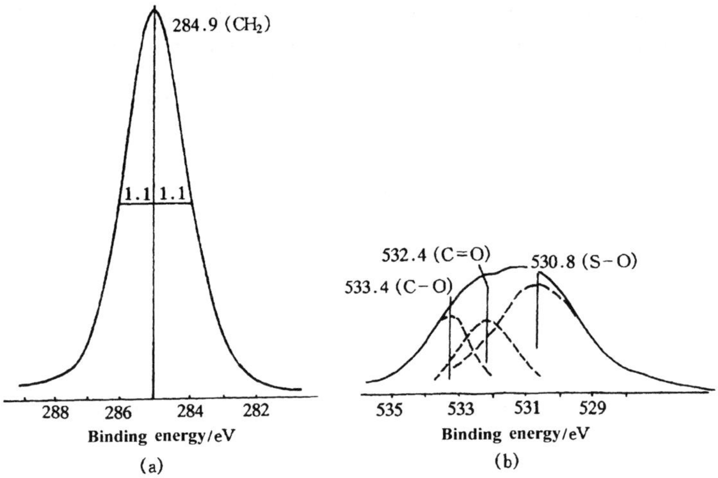

Fig.8.2. XPS spectra of worn surface of natural rubber eroded in PAM
(a) Carbon element (C_{1S}); (b) Oxygen element (O_{1S}).

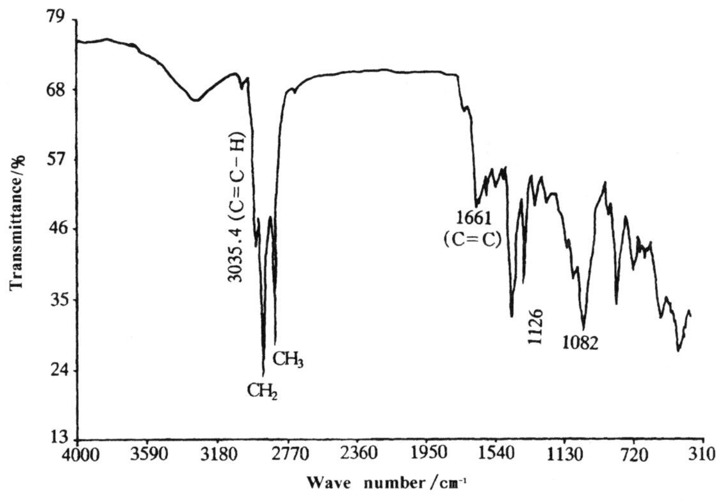

Fig.8.3. FTIR spectrum of the original surface of natural rubber

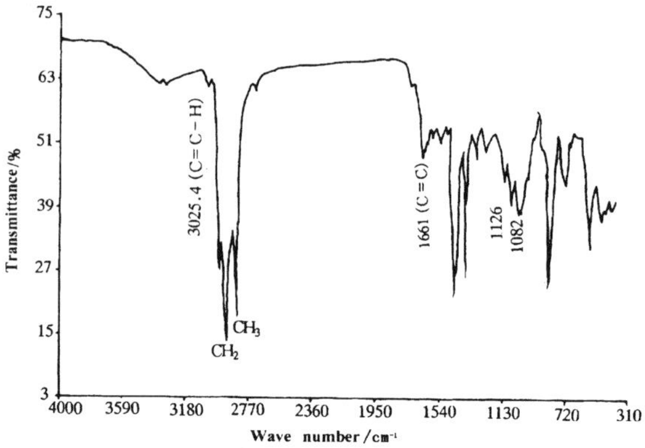

Fig.8.4. FTIR spectrum of the worn surface of natural rubber eroded in PAM solution

Under the action of applied force resulted from the impacting, scratching, rolling and sliding of the flowing quartz sand with a certain amount of kinetic energy, there is no time for the macromolecular chains of the rubber surfaces to relax. Thus, the stress waves propagate forwardly and continually, and maintain the acute front-peaks. Therefore, more strong stress field exists in the surfacial layer of nature rubber and makes the molecular-chain segment of surfacial layer in a very small volume to be exerted by a larger stress. When the stress is large enough, the increment of bond-angle deformation of the covalent bonds within the molecular chains results in the increase in chemical activity of organic covalent bonds, such as -C-C-C-, -C=C- and -S-S-C--, and the macromolecular chains are excessively activated. Thus the strength and energy of covalent bonds are decreased under the extension action of the particles on the molecular chains of the rubber. In the meantime, the reactivity and the atomic distance of covalent bonds are increased. As a result, the potential energy between the covalent bonds is re-allocated [152], and the activation energy of chemical reactions decreases. At this moment, the rupture of molecular chains of the rubber occurs and the primary active free-radicals are produced.

The fracture position of macromolecular chains depends on their structure and the intensity of stress being applied on the molecular-chain segments [153]. It might be located at the weakest bond (α -methylene), and then the free-radicals of poly-iso-2-pentene are generated.

The process of fracture is schematically given as follows:

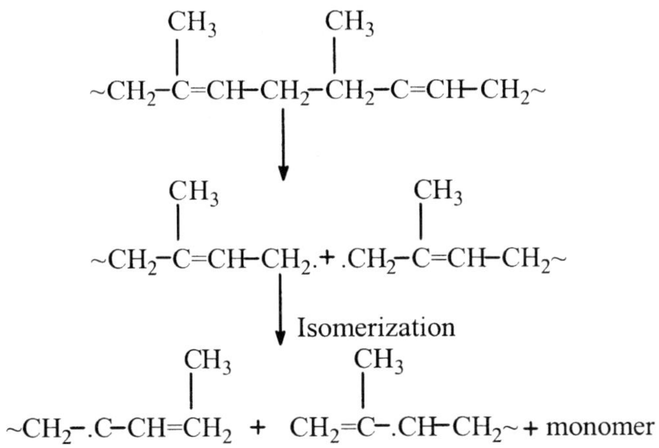

When the ruptured free-radicals are re-allocated, the macromolecular chains react with the adjacent primary free-radicals of poly-iso-2-pentene, and then induce chain branching and cross linking. The process of reaction of branching is given by:

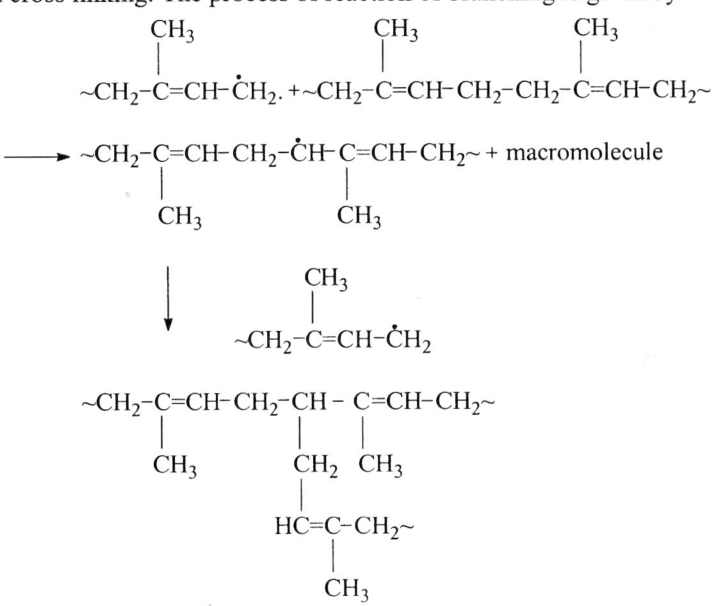

Besides the above branched and crosslinked reactions between the more activated free-radicals and the macromolecular chains, the activated free-radicals are apt to react with oxygen in the absorption layer and also resulted in the macromolecular chains being branched and crosslinked. The chorionic membrane existing on the worn surface of nature rubber, as described in Chapter 6, might be the branched reaction products as the branched chains of alkene are generally existed on the surface.

The molecular weight for the fractural fragment of macromolecular chains of nature rubber cannot less than a critical value M_{∞}, namely the molecular weight of the shortest ruptured chain segment in theory [153,154].

$$M_{\infty} = E_1 m / E_2 \quad (8\text{-}1)$$

Where, E_1-activation energy ruptured of covalent bonds in main chains;

m - molecular weight of ruptured chain element;

E_2 - activation energy to conquer the interactions between moleculae of each chain segment.

The free-radical termination reactions between macroradicals and the nucleophilic addition reactions of free radicals between macroradicals and -C=C- of molecular chains might occur, which can produce another crosslinked network structure. Thus, the surfacial mechanical properties of the rubber are become so brittle that the wear-resistance of rubber is reduced. This is one of the reasons of nature rubber being easily aged.

Due to the mechanical action of cohering particles in PAM solution on the rubber surface, a great number of crosslinked structures of covalent bonds in the macromolecular chains, namely, -S-S- crosslinked bond, are ruptured. As a result, partial network structures of NR materials are destroyed, and the free-radicals containing element of sulphur are produced. As these free-radicals is apt to react with the oxygen in water-absorbing layer, the covalent bonds of sulphur-oxygen (-S-O-) are formed (Fig.8.2b). The process of reaction is given by:

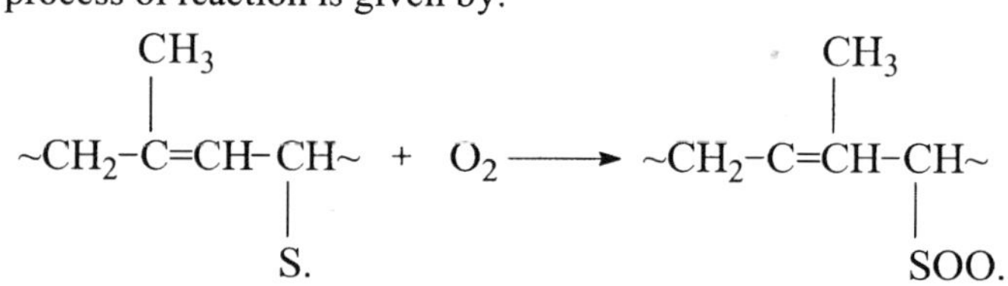

The rupture of crosslink results in the decrease in the degree of crosslinking and the increase in the chain flexibility. Therefore, the molecular chains are easily to be cut off by the particles, and a number of primary active macroradicals are emerged.

During the abrasive erosion, the fracture of macromolecular chains of PAM solution itself also occurs due to the action of particles. The fracture position is located at the site of –C–C–, namely:

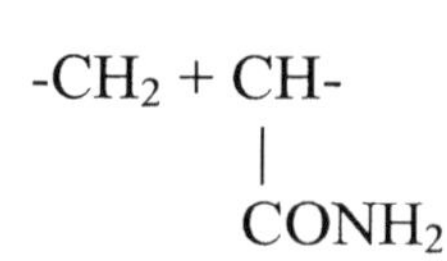

Therefore, the flowing active free-radical groups of polyacrylamid exist in the solution and might be combined with the free-radicals of the rubber surface. Thus, the PAM solution itself becomes a recipient of active free-radicals of NR and intensifies the fracture of macromolecular chains on the worn surface of the rubber eroded in PAM. Hence, the network structure with high molecular weight is yield, which reduces the flexibility of the molecular chains and causes the material to be brittle. As a result, peeling occurs on the rubber surface, which results speedily in the wear of material. This is one of the reasons for the corrosion hollows produced on the surface of NR material (Fig.6.9). It could be proved by the element of nitrogen existing on the worn surface.

8.2.2 Surface Oxygenated Degradation

From the photoelectron peaks of energy levels of the carbon elements (C_{1S}), it could be found that the sort and amount of functional groups containing oxygen on the worn surfaces are more than those on the original surface. As shown in Figures 8.2, 8.5 and 8.6, on the worn surfaces eroded respectively in three different media, there are existed functional groups of C=O, -C-O- and -COOH, etc. Moreover, based on the ferrographic analysis, it has been revealed that the groups of -S-O- and RCOONa occur on the worn surface eroded in PAM and in NaOH respectively (Fig.8.2 and Fig.8.6). It could be inferred from the emergence of the above groups that the surfacial oxygenated degradation is generated on the rubber surface during eroding as it is impossible to be hydrolysis for nature rubber eroded water due to the character of its structure of molecular chains [154].

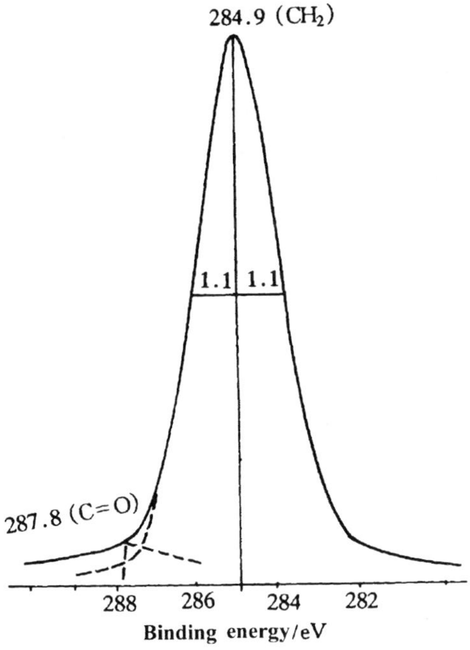

Fig.8.5. XPS spectrum of the spectral peak of carbon element in the worn surface of NR eroded in H_2O

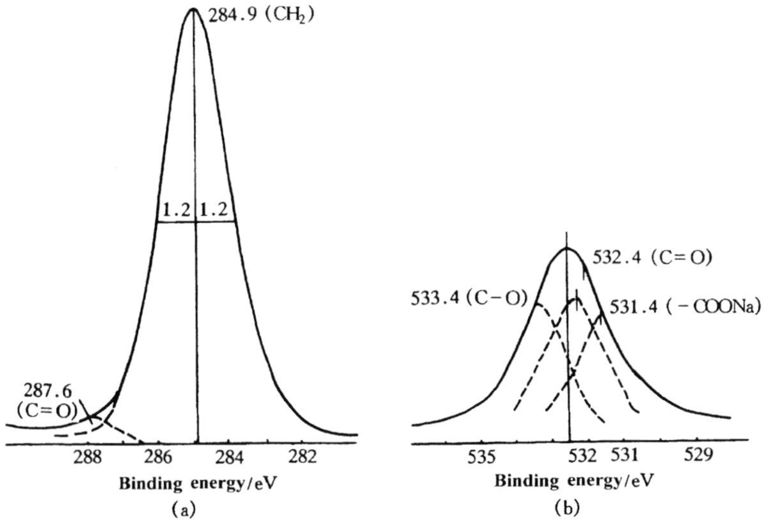

Fig.8.6. XPS spectrum of worn surface of NR eroded in NaOH solution
(a)Carbon element (C_{1S}); (b) Oxygen element (O_{1S})

It is also found that some new groups containing oxygen are produced as the peak areas of functional groups containing oxygen increase and the peak of carbonyl groups (-C=O) is widened on the basis of FT-IR analysis (Fig.8.7). Moreover, the obvious peaks of carboxyl groups (-COOH) are also observed on the worn surfaces eroded respectively in media of PAM and NaOH, and the amount of carbonxyl groups in NaOH solution is greater than that in PAM solution. However, the carbonyl groups are the most in medium of PAM. In addition, the peak area of -C=C- on the worn surfaces decreases and its wavenumber is higher than that of the original surface as the groups of -C=C- are oxidized, which also proves the occurrence of surfacial oxidation.

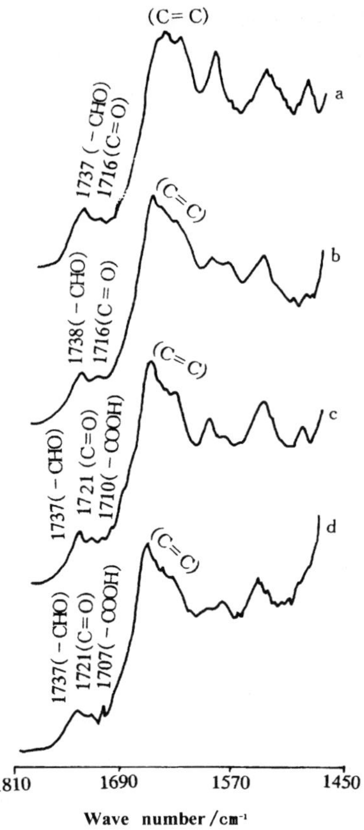

Fig.8.7. FTIR spectra of NB (1450~1810cm^{-1})

a – original surface; b – worn surface in H_2O;

c – worn surface in PAM; d – worn surface in NaOH

The main products of oxygenated degradation of macromolecular chains in the surface layer are polymers with terminal groups containing oxygen and the other low-molecular weight substances, which are located at the rupture site of chains.

Under the repeated action of impacting, scratching, rolling, rubbing of particles in liquid medium, the hysteresis deformation occurs many times and generates a large amount of accumulated heat at a certain depth beneath the surface as nature rubber is a high-elastic material, which provides much enough energy for oxidation reaction of the macromolecular chains. A great number of primary activated macroradicals are produced due to the rupture of molecular chains of the rubber surface. Although some macroradicals are flushed away by the fluid media, there is still quite a number of activated free-radicals retained on the worn surface while the process of fracture of macromolecular chains is proceeding continuously during abrasive erosion. In addition, the adsorbed water layer on the rubber surface provides enough oxygen for the oxidation of the molecular chains of NR. Therefore, the activated free-radicals on the rubber surface are apt to react with oxygen. The sticky layer formed on the worn surface of nature rubber is the very result of oxygenated degradation of the molecular chains of the rubber [56,154].

The main mechanism of that activated free-radicals reacts with oxygen is the chain polymerization reaction of free-radicals with self-catalysis. The peroxide free-radicals are produced by the reaction of activated radicals to oxygen in water. The process of oxygenated degradation is given as follows:

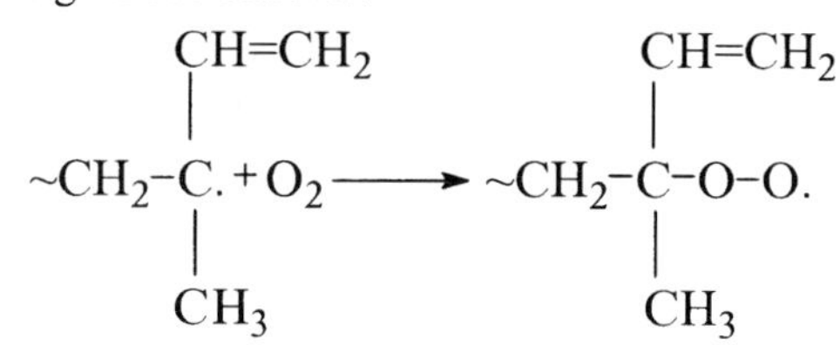

When the peroxide free-radicals reacts with the adjacent macromolecular chains of nature rubber and captures their hydrogen atoms, the hydrogen peroxide is produced, that is, R-COO·+RH⟶ RCOOH+R·.

The life span of hydrogen peroxide being longer is one of the reasons for the occurrence of the single covalent bonds of carbon-oxygen on the worn surfaces. However, due to the absorbed water layer existing on the surface during abrasive erosion, somc hydrogen peroxides are decomposed, that is, RCOOH⟶ RO·+·OH. Then the functional groups of RO· might be resolved further, which results in fracture of macromolecular chains, namely the molecular rearrangement of chains parts occurs. By means of a series of reactions and transmission, RCOOH turns into oxidative products containing functional groups of carbonyl groups (-C=O). The process of these reactions is given as follows:

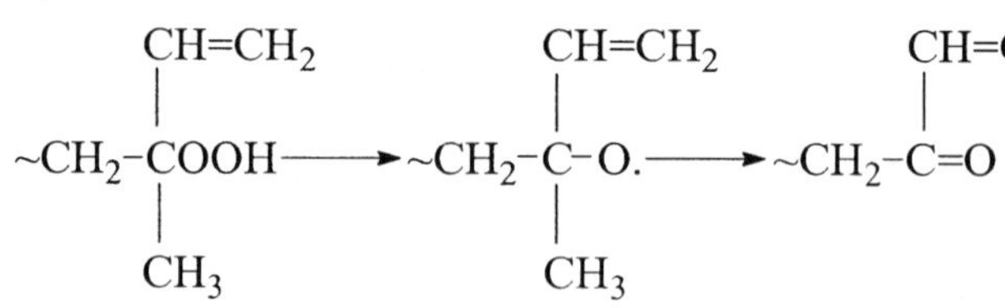

I II

The chains are branched when the peroxide free-radicals react with adjacent macromolecular chains by capturing their hydrogen and attacking the -C=C- of the macromolecular chains. It is also one of the reasons for nature rubber being apt to age [152]. The existing of downiness substance on the worn surface observed by SEM is possibly the expression of these chains being branched.

The oxygenated degradation also might occur when the oxygen reacts with the groups of -C=C- on the macromolecular chains under the mechanical action of particles. The reaction processes are given as follows:

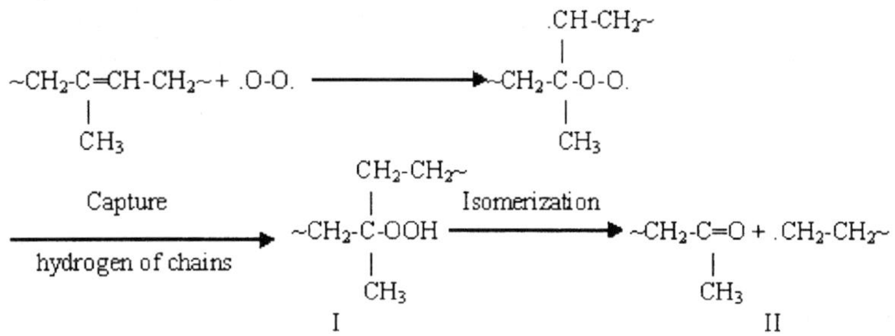

In summary, by means of a series of oxygenated degradation for free-radicals of macromolecular chains, the low molecular weight oxidative products containing functional groups of aldehyde and ketone with branched chains are produced. The process of reaction is given as follows:

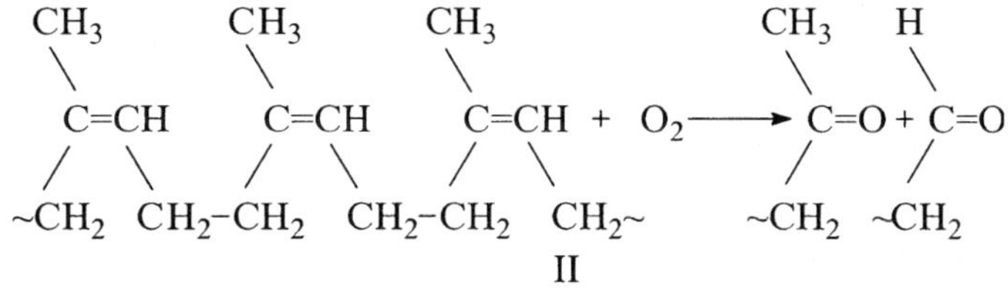

The functional groups, including carboxyl (-C(=O)-OH) and aldehyde (-C(=O)-H), were found on the worn surface by using XPS and FTIR. They are the products of -C=O being oxidated further.

Therefore, the occurrence of sticky layer on the worn surfaces of nature rubber (Figures 6.9, 6.10 and 6.11) is resulted from the existence of functional groups, such as, carbonyl (aldehyde and ketone) and hydrogen peroxide.

Owing to the eroding of flowing fluid, the oxidative macromolecular chains of nature rubber on worn surfaces are flushed away continually and resulted in the loss of materials.

In addition, the α -methyl on the macromolecular chains of nature rubber might be also oxygenated. As the special location of α -methyl, it has intensive activation under the influences of adjacent groups of -C=C- and methylenes. Moreover, the activated α -methyl could be oxygenated when heat resulted from the inner friction of rubber is accumulated beneath the worn surface. The main oxidized products are the polymers containing carboxyl and some substances with low molecular weight.

By the FTIR analysis (Fig.8.7), it has been found that the main oxidized products on the worn surface of nature rubber eroded in three different media are I (C-O) and II

(C=O). Besides, the carboxyl groups are found on the worn surface eroded in PAM or in NaOH. Therefore, the extent of oxygenated degradation for macromolecular chains eroded in PAM or in NaOH respectively is larger than that in H_2O. Owing to the number of oxidized products containing carboxyl groups on worn surface eroded in NaOH is higher than that in PAM, the reaction of oxygenated degradation in NaOH is most obvious and throughout among the three media, which results in the most loss of material. From above, it shows that the oxygenated degradation of nature rubber eroded in PAM and in NaOH respectively has some specialties except for the common features.

The proportion of single covalent bonds of carbon-oxygen on the worn surface eroded in PAM is the most among the three different media. The main products of oxygenated degradation for nature rubber are the monomers and the low molecular weight polymers with -C-O- terminal groups.

Under the experimental condition, the hydrolysis of polyacrylamide is occurred as shown by

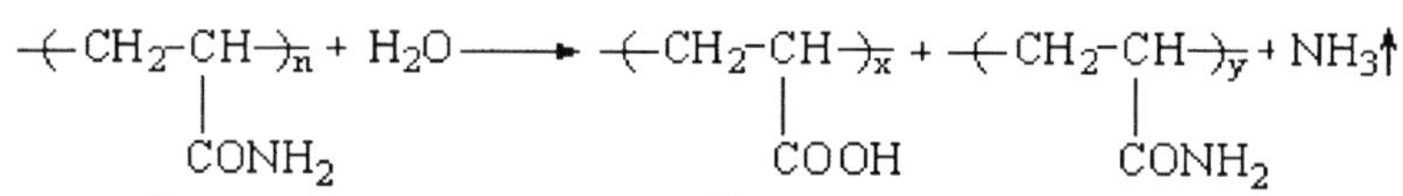

I II III (1)

Due to the polyacrylamide solution being exhibited acidic (pH=6), the polyacrylamide and the hydrolzate (II) are analogous to an indirect and a direct catalyst respectively. Therefore, the oxygenated degradation of nature rubber eroded in this medium is promoted. However, it is easy to produce electrophilic addition reaction between -C=C- of macromolecular chains and H_2O. The process of reaction is given as follows:

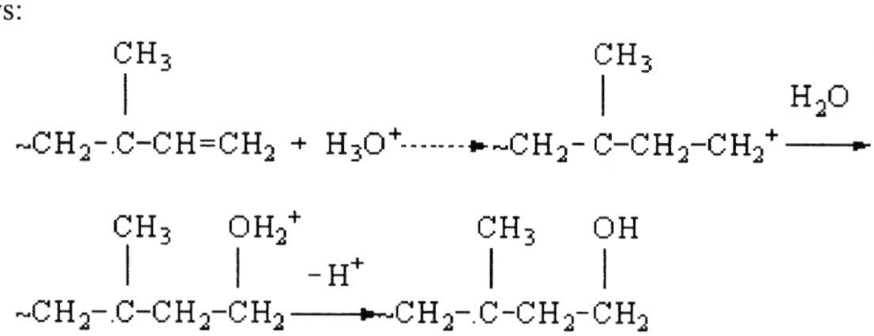

IV (2)

The products (IV) containing free-radicals with hydroxyl groups on the chains are able to react with oxygen in absorbed water layer further, and then produce the groups of aldehyde $(-\overset{\overset{O}{\|}}{C}-H)$ and ketone $(-\overset{\overset{O}{\|}}{C}-)$ on the branched chains. As a result, the amount of groups of C=O and C-O on worn surface eroded in PAM is the most. In the meantime, some reactive products flush away, namely, the surface is worn out. Thus, the reaction is preceded to the right of formula (2) based on the theory of dynamic equilibrium of reversible chemical reactions. Therefore, the reactive substances in the left of formula (2) are reduced unceasingly, and the products (IV) increase. These processes proceed continuously, and then cause the corrosive hollows to be occurred. Thus, the wear value of the rubber eroded in PAM is higher than that in H_2O.

As stated above, the mechanism of oxygenated degradation of nature rubber in PAM is more complicated than that in H_2O. In order to reveal this mechanism completely, there is still much work to be done.

The amount of carboxyl groups and the decreasement of methylene groups (-CH_2-) on the worn surface of NR in NaOH are the most by comparison with these in the other two media (Fig.8.7), which show that the oxygenated degradation is the main surfacial mechanochemical reaction of nature rubber eroded in NaOH. Moreover, during the process of oxygenated degradation, the wear value of nature rubber is large (Table 6.1) and the corrosive hollows occur (Fig.6.10) as the NaOH solution is acted as the catalyst and reaction substances. These conclusions can be proved by the existence of sodium and -COONa groups on the worn surface (Fig.8.6).

The process of oxygenated degradation of nature rubber eroded in NaOH is given as follows:

$$\sim CH_2-\underset{}{\overset{\displaystyle OH}{\overset{|}{C}}}=O + NaOH \longrightarrow \sim CH_2-\overset{\displaystyle ONa}{\overset{|}{C}}=O$$

In order to constrain effectively the oxygenated degradation of nature rubber eroded in the three different media, a useful way to end the succession reaction of free-radical is adding the stabilizers during the process of the rubber being pressed into shape. These stabilizers are some of phenol category with stronger potential energy, including derivative of phenol or cresol, which are apt to react with activation free-radical. This is a method most in use to prolong the life time of nature rubber nowadays.

8.3 Surfacial Mechanochemical Effects of Nitrile Rubber

Nitrile rubber (NBR) is copolymerised of butadiene and acrylonitrile. Its property is mainly depended on the acrylonitrile as the content of acrylonitrile is up to 34%. The structural formula of nitrile rubber is given as follows:

$$-[(-CH_2-CH=CH-CH_2-)_x-(-CH_2-\underset{\displaystyle CN}{\underset{|}{CH}}-)_y]_n \qquad (n=1,2,3,...)$$

The oil-resistance and tolerance to polarized liquid of NBR are much better. Moreover, its heat-resistance, wearability, corrosion-resistance and air-tightness are better than that of nature rubber. Although nitrile rubber could be used under 120℃ for a long time, its cold-resistance is worse.

XPS analysis of the rubber surfaces is performed in order to disclose the surfacial mechanochemical changes during abrasive erosion. The elements and binding energies for the original and worn surfaces are shown in Table 8.1.

Table8.1. The results of XPS analysis of nitrile rubber

Specimen	Elements composition(wt%)						Binding energy(eV)			
surface	C	O	N	S	Si	Na	C1s	O1s	N1s	
Original surface	89.8	5.5	1.7	1.9	1.0	/	285.0	532.2	399.6	168.7
Worn surface (in H2O)	79.4	15.2	2.0	2.0	1.4	/	285.0	532.3.	400.5	168.8
Worn surface (in PAM)	86.6	10.0	2.1	0.3	1.1	/	285.0	533.1	400.9	169.4
Worn surface (in NaOH)	71.4	16.6	4.6	0.7	4.0	2.6	285.0	532.3	400.2	168.7

Based on the XPS analysis, it could be deduced that three categories of surfacial mechanochemical reactions occur on the worn surfaces eroded in three different media respectively.

8.3.1 Fracture of Macromolecular Chains

The decrement of weight concentration (wt%) of carbon of the worn surfaces shows that the macromolecular chains are fractured and then the free-radicals are produced (Table 8.1). Moreover, a number of crosslinking covalent bonds of S-S might be ruptured on the worn surface of the rubber eroded in PAM, which is probably accounted for the almost disappearing of the sulphur on the worn surface (Table 8.1).

In a similar way to the rupture of natural rubber, the fracture of macromolecular chains is also emerged on the surfaces and subsurfaces of nitrile rubber under the action of impacting, scratching, rolling and rubbing of particles with a certain amount of kinetic energy.

As known from the physical process of the abrasive erosion of nitrile rubber as mentioned in Chapter 6, the fracture of chains on subsurface is caused by the action of the maximum shear stress field, and that on the surface is probably resulted from the maximum thermal stress field. Therefore, some free-radicals, low weight molecular substances, monomers and new crosslinked network structure are produced.

The processes of fracture of macromolecular chains are given as follows:

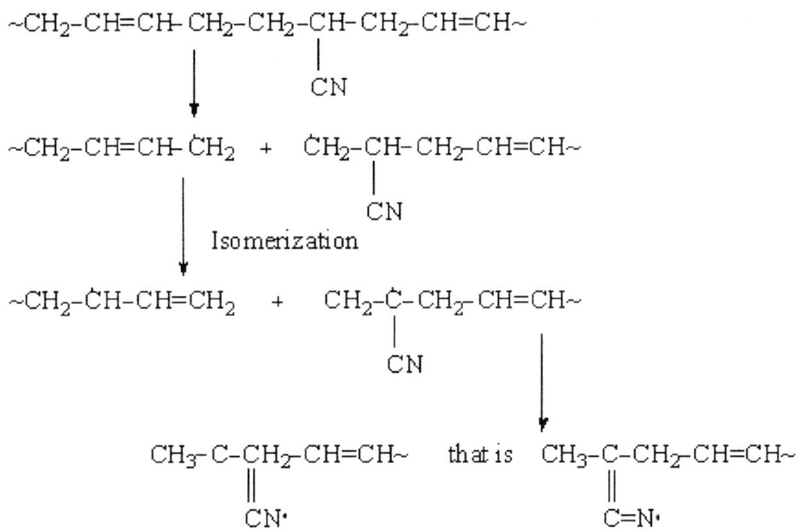

During the process of isomerization of chains, some low weight molecular substances are also generated. As the free-radicals of NBR have a certain activity, the reactions of both chain transfer and chain termination could be produced. When nitrile rubber is eroded in the media of H_2O or PAM, the free-radicals on subsurfaces transfer to adjacent macromolecular chains and become stable end groups, which might cause the rubber to produce micro-cracks. Moreover, the reactions among the active free-radicals are able to form new crosslinked network structure and result in the rupture of macromolecular chains. Therefore, the fatigue wear is the main type of wear for the nitrile rubber eroded in water or in polyacrylamide solution (Figs. 6.12 and 6.13). Owing to the degree of the hydrolysis of macromolecular chains eroded in NaOH being the most, the mechanical properties of NBR reduce, and then the surface of rubber is worn out speedily under the microcutting of particles and the eroding of liquid. As a result, there is no time for the microcracks to be propagated. Therefore, the main mechanism of wear is microcutting, which is probably accounted for that the wear of NBR eroded in this medium is the most comparing with the other two media.

By comparison with nature rubber, the macromolecular chains of nitrile rubber are more prone to be cut off [154] and the free-radicals are more apt to initiate reactions by other free-radicals. Therefore, the number of fracture of macromolecular chains and the wear value of nitrile rubber are greater than that of nature rubber.

8.3.2 Surface Oxygenated Degradation

From the XPS analysis, the functional groups containing oxygen, such as, RCOOH and R-OH, are found on the worn surface and the intensity of higher-binding energy region of carbon element (C_{1S}) levels is much stronger than that of original surfaces.

These phenomena indicate that the oxygenated degradation occurs. By comparison with nature rubber, the oxygenated degradation for NBR eroded in the three kinds of media is weaker because the number of groups of -C=C- is minor, moreover, the activation of -C=C- and the free-radicals is confined by the CN groups [152]. However, the mechanism of oxygenated degradation of NBR is still the chain polymerization reaction of free-radicals with self-catalysis similar to that of NR. And also, when nitrile rubber is eroded in PAM, the electrophilic addition reaction between carbon-carbon double bonds(C=C) and H_2O occurred in the same way as that for NR. Therefore, the degree of oxygenation and the binding energy of oxygen on worn surface eroded in this medium are the most (Table 8.1).

The process of electrophilic addition reaction between carbon-carbon double bonds and H_2O is given as follows:

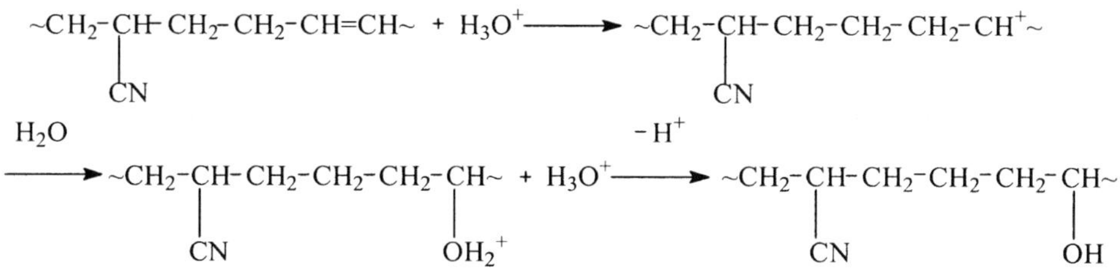

I

Similar to NR, the peroxide groups are produced when free-radicals react with the oxygen in absorption layer, and the hydrogen peroxide groups are formed whilst the free-radicals react with the adjacent macromolccular chains. Although the hydrogen peroxide groups are more stable, the isomerization reaction might occur, which causes the oxygenated chains to rupture. As a result, the oxidates including hydroxyl (-R-OH) and carboxyl (-R-COOH), are produced. This process is also proved by the increase in oxygen content on the worn surfaces (Table 8.1).

The process of reactions is given as follows:

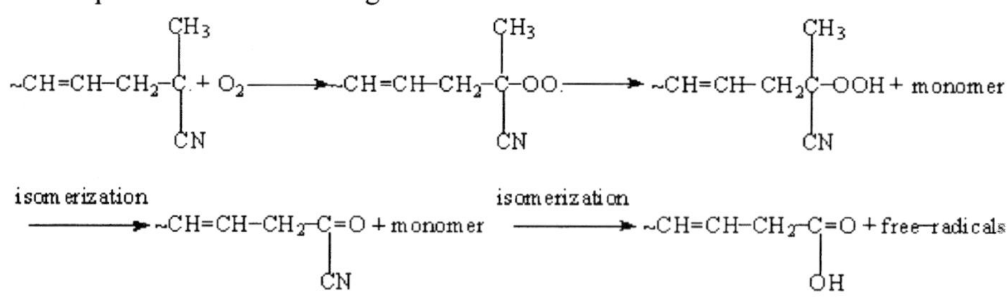

Much the same as the nature rubber, the carbon-carbon double bonds could be oxygenated by oxygen absorbed in the water layer. As the existing of carbon positive ion, some by-products are produced in the reaction process. A small amount of the product I might be further oxygenated and become carboxyl groups.

Since the free-radicals of nature rubber are more active than that of nitrile rubber, the degree of oxygenated degradation of the former is more complete than that of the latter. Therefore, oxygenated degradation is not the main surfacial mechanochemical effect of nitrile rubber eroded in the three different media.

8.3.3 Hydrolysis

Analyzing the worn surfaces of nitrile rubber eroded in the three different media respectively, it can be observed that the functional groups contained oxygen, including acylamine groups (-$CONH_2$) and carboxyl groups (-COOH), are generated (Fig.8.8, Fig.8.9 and Fig.8.10). In addition, the carboxylic acid sodium groups (-COONa) are found on the worn surface eroded in NaOH (Fig.8.10). Moreover, the amount of carboxyl groups (-COOH) on worn surfaces eroded in PAM is more than that of in H_2O (Fig.8.8 and Fig.8.9). Therefore, it could be deduced that the hydrolysis occurs on the worn surfaces. Most of carboxyl groups (-COOH) are probably hydrolysis products of acylamine groups (-$CONH_2$), except a small amount of products of oxygenated reaction. As the -CN groups on the macromolecular chains of NBR have strong polar and are apt to be hydrolyzed, the hydrolysis is the main surfacial mechanochemical effect of nitrile rubber eroded in the three different media respectively. The extent of hydrolysis of NBR in the three kinds of medium is in turn from strong to weak as follows: NaOH>PAM>H_2O. The above deductions are also verified by the fact that the relative oxygen-content (%) of the worn surfaces is higher than that of the original surfaces (Table 8.1).

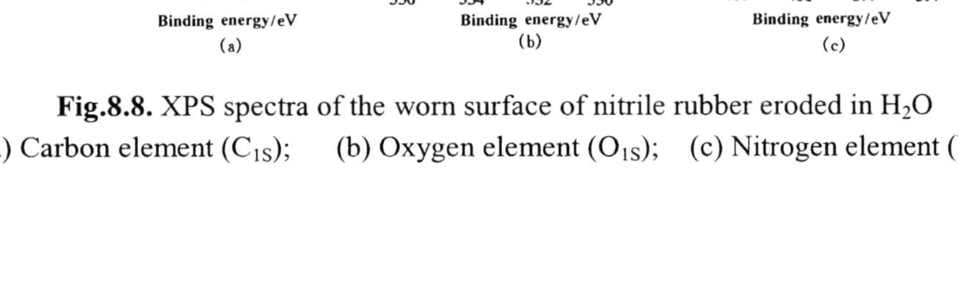

Fig.8.8. XPS spectra of the worn surface of nitrile rubber eroded in H_2O
(a) Carbon element (C_{1S}); (b) Oxygen element (O_{1S}); (c) Nitrogen element (N_{1S})

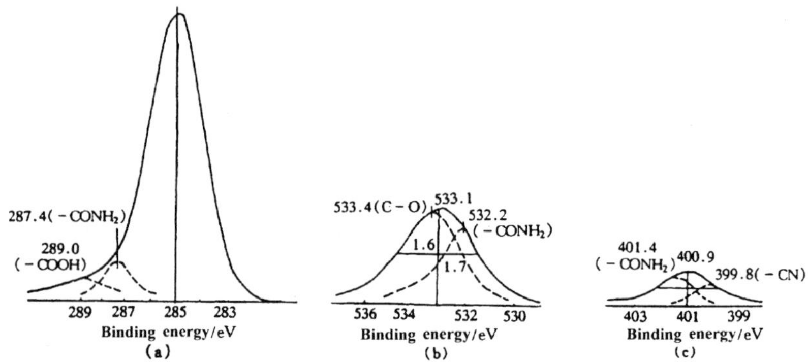

Fig.8.9. XPS spectra of the worn surface of nitrile rubber eroded in PAM
(a) Carbon element (C_{1S}); (b) Oxygen element (O_{1S}); (c) Nitrogen element (N_{1S})

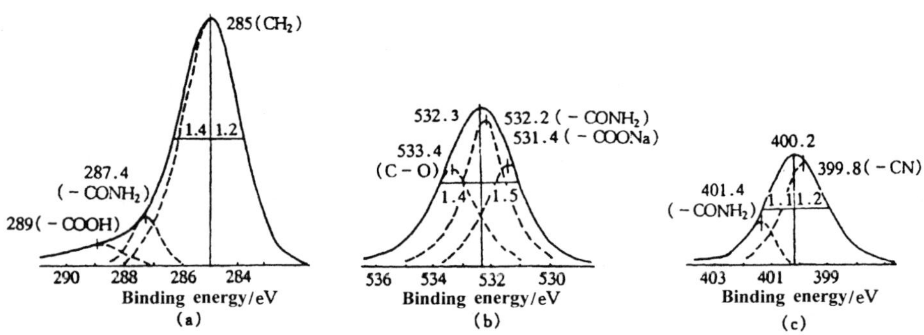

Fig.8.10. XPS spectra of the worn surface of nitrile rubber eroded in NaOH
(a) Carbon element (C_{1S}); (b) Oxygen element (O_{1S}); (c) Nitrogen element (N_{1S})

Owing to the action of impacting, rolling and rubbing of the particles as well as the temperature effect, the macromolecular chains on the surface of nitrile rubber are much activated. Therefore, the -CN groups on chains can react with H_2O, and then result in hydrolysis. The products of hydrolysis are acylamino-groups (-$CONH_2$).

The process of hydrolysis is given as follows:

$$H_2O \longrightarrow H^+ + OH^-$$

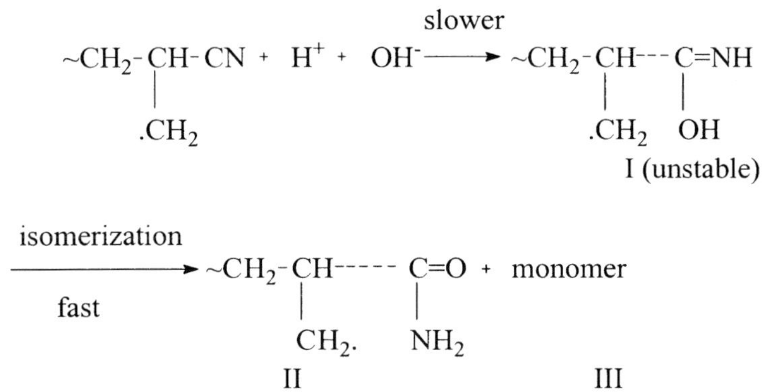

The products (II) and (III) are easily to be flushed away. Under the influence of NH_2 groups on chains, the binding energy of -C=O in acylamin $(-\overset{O}{\overset{\|}{C}}-NH_2)$ is less than that in aldehyde and ketone as the electronic luring effect of nitrogen is weaker. Owing to the products being flushed away continually, the above reaction is preceded in the direction of right. Thus, a large amount of wear of materials is produced.

Due to the PAM solution being acidic (pH=6), the extent of hydrolysis of -CN groups in this medium is larger than that in H_2O as the -CN groups are prone to be hydrolyzed under the condition of sour solution. Possibly, partial products (II) containing acylamin are also hydrolyzed further to produce carboxyl groups. This reaction in PAM is more obvious than that in H_2O (Fig.8.8a and Fig.8.9a). The reaction process is given as follows:

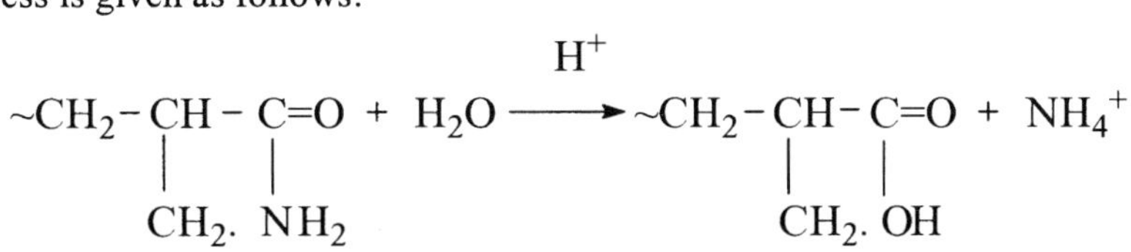

It has been found that the carboxylic acid sodium groups (-COONa) are existed on the worn surface (Table8.1 and Fig.8.8b). As the NaOH solution exhibits basic (pH=12), the degree of hydrolysis of -CN groups in this medium is the most by comparison with that in PAM and in H_2O. A great deal of carboxyl groups (-COOH) are produced and then reacted with NaOH. Therefore, during the process of hydrolysis of nitrile rubber eroded in NaOH, the medium becomes both catalyst and reaction substance. As a result, the wear value of NBR eroded in this medium is the most relative to that in PAM and in H_2O. The reaction process of -CN groups with NaOH solution is given as follows:

$$NaOH \longrightarrow Na^+ + OH^-$$
$$H_2O \longrightarrow H^+ + OH^-$$

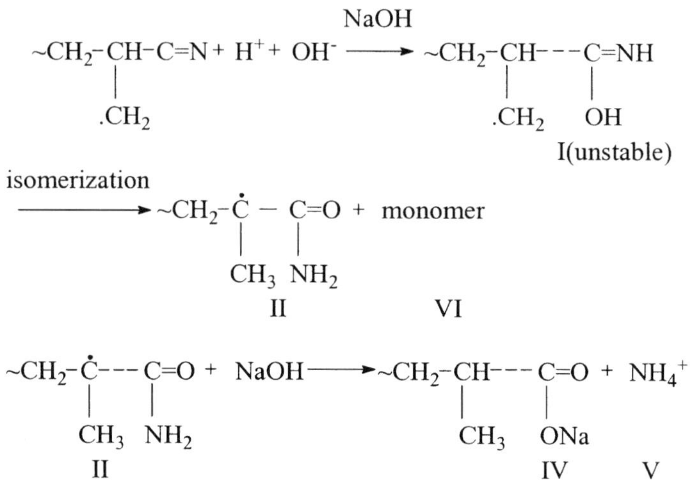

The reaction products (IV), (VI) and (V) being flushed away continually cause the rubber to wear out incessantly. The above reaction process are proved by the changes of oxygen-content and sodium-content on worn surface (being 16.6% and 2.6% respectively) as well as the existence of carboxylic acid sodium (-COONa) on the worn surface (Table 8.1 and Fig.8.10b). Since the severe extent of hydrolysis of NBR in NaOH is the most among the three media, no fatigue and delamination occurs on the worn surface (Fig.6.14).

8.4 Surfacial Mechanochemical Effects of Fluororubber

Fluororubber is a copolymer of crystallized vinylidene fluoride monomer (UDF) and other monomers containing asymmetric fluorine element. The structural unit of molecular chains is shown as follows:

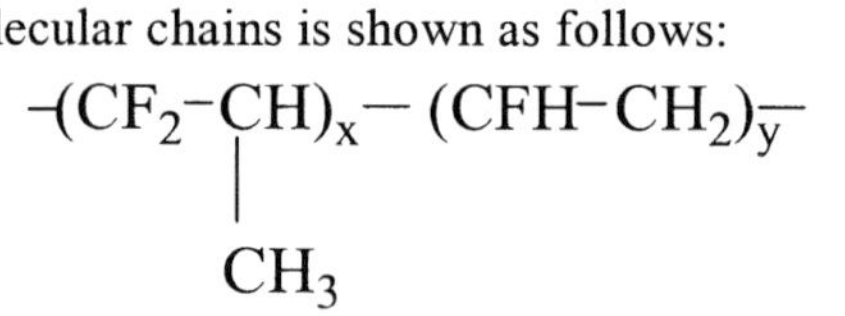

(x=1,2,3,....; y=1,2,3,....)

The heat-resistance of fluororubber is so strong that it can be used under the condition of high temperature up to 300℃. Moreover, its oil-resistance, corrosion-resistance and oxidation resistance are better. As for its tear resistance, it is similar to that of natural rubber. Although fluororubber is a kind of special synthetic rubber with better comprehensive properties, its cold-resistance and venting quality are worse.

8.4.1 The Fracture of Macromolecular Chains

Similar to the fracture of macromolecular of NB and NBR materials, the free-radicals are produced when the macromolecular chains of fluororubber are ruptured. As the free-radicals react with the adjacent chains, some new crosslinked structures are generated.

From the FT-IR spectra (Fig.8.11 and Fig.8.12), it is found that the peaks of functional groups of -CH, -CH_2 and -CF are reduced on the worn surfaces. Moreover, the peak of CH on the worn surfaces eroded respectively in PAM and in NaOH is disappeared. Therefore, it shows that the carbon-carbon cross linking might be formed after the rupture of molecular chains.

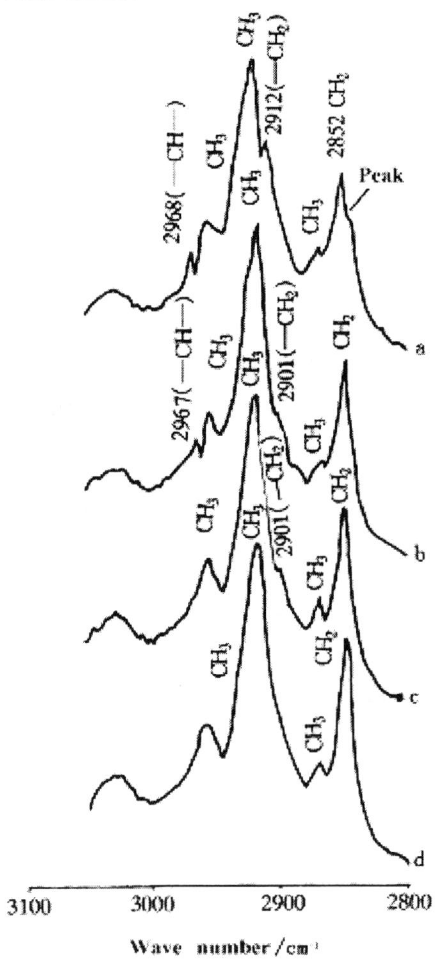

Fig.8.11. FTIR spectra of fluororubber (3100~2800cm^{-1})

(a) Original surface; (b) Worn surface (in H_2O); (c) Worn surface (in PAM);

(d) Worn surface (in NaOH)

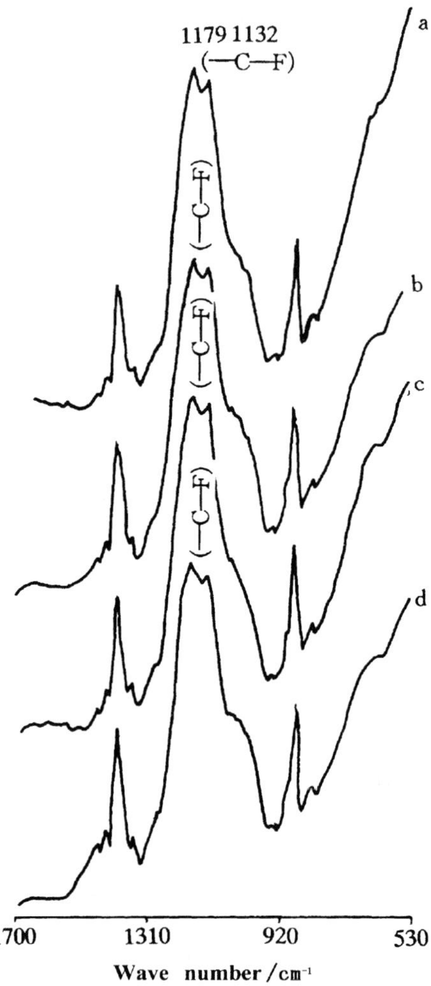

Fig.8.12. FTIR spectra of fluororubber (1700~530cm^{-1})
(a) Original surface; (b) Worn surface (in H_2O); (c) Worn surface (in PAM); (d) Worn surface (in NaOH)

The active free-radicals stemmed from the rupture of molecular chains are able to be captured by the adjacent molecular chains of fluororubber and reacted with the hydrogen and fluorine of functional groups, including -CH, -CF_2 and –CF. Thus, the cross linking reaction occurs. It results in the increase in the surfacial density, and then the physical properties of rubber surface are worsened. The reaction process is given as follows:

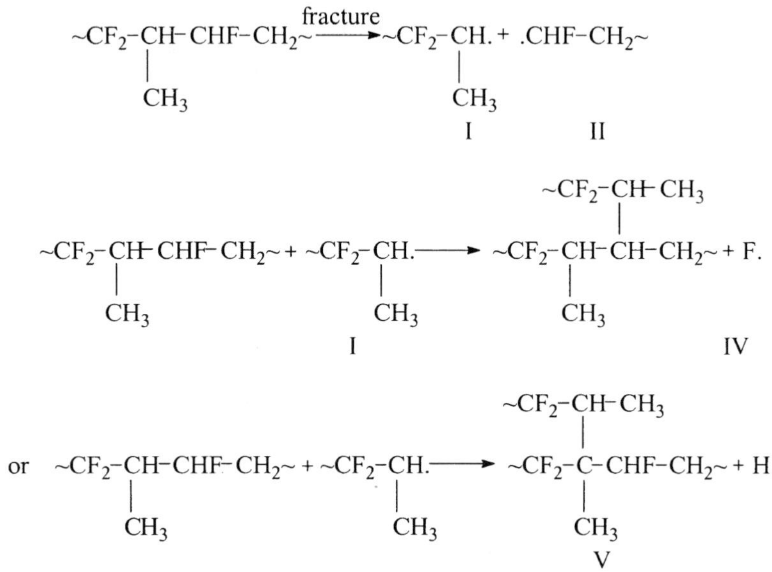

The molecular chains of fluororubber can not produce small monomer after fracture as the covalent bonds of carbon-fluorine existed in the chains are much stronger, which is different from that of nature rubber and nitrile rubber.

The fracture of macromolecular chains would be the main surfacial mechanochemical effects of fluororubber eroded in H_2O or in PAM.

Due to the action of cohering particles of PAM solution, the interaction between the rubber surface and the flowing particles is so strong that the macromolecular chains on the surface is in a much intensive excited state, which might induce a large number of active free-radicals groups and hydrogen free-radicals (H.). Thus, the degree of cross-linking is obviously raised and the mechanical properties of fluororubber surface decreases. Therefore, the wear value of the fluororubber eroded in PAM is greater than that in H_2O.

As for the rubber surface eroded in NaOH, its main surfacial mechanochemical effect is not the fracture of macromolecular chains because a large number of carbonyl groups of C=O were found on the worn surface by using FT-IR (Fig.8.13).

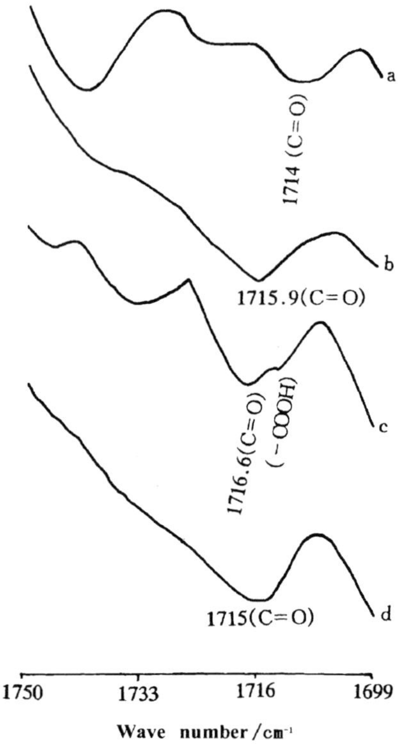

Fig.8.13. FTIR spectra of fluororubber (1750~1699cm-1)
(a) Original surface; (b) Worn surface (in H_2O); (c) Worn surface (in PAM)
(d) Worn surface (in NaOH)

8.4.2 Surface Oxygenated Degradation

By using FT-IR, the carbonyl groups (-C=O) on worn surfaces of fluororubber have been identified. As seen from Fig. 8.13, the peak of carbonyl groups (-C=O, 1715.9cm^{-1}) on the worn surface eroded in H_2O is found (Fig.8.13, b). As for the peak of carbonyl groups (-C=O, 1716.6cm^{-1}) on the worn surface eroded in PAM, its site is higher and its area is larger by comparison with the above peak of carbonyl groups. Moreover, a small peak of arboxyl groups (-COOH) can be observed beside it (Fig.8.13, c). These phenomena shows that the sorts of functional groups containing oxygen on the worn surface eroded in PAM are more than that in H_2O. On the worn surface eroded in NaOH, the area of the single peak of carbonyl groups (-C=O, 1715.0cm^{-1}) is the most in comparison to the worn surfaces eroded respectively in the other two media (Fig.8.13, d). Therefore, the amount of groups occurring on the worn surface eroded in three media respectively is put in order from large to small as follows: NaOH >PAM >H_2O

(Fig.8.13). Obviously, the degrees of oxygenated degradation on the rubber surface eroded in three media respectively are also in the same order. The oxygenated products are mainly macromolecular chains containing groups of C=O.

Based on the XPS analysis, some functional groups containing element of oxygen, including CF-O, C=O and C-O, have been found on the worn surface eroded in PAM (Fig.8.14). On the worn surface eroded in NaOH, some functional groups containing oxygen occurred at the positions of corresponding groups, such as, C-C(C=O, C-O), CF(C=O) and CF_2 (CF-O), have been identified (Fig.8.15). The results of XPS analysis of the rubber surfaces before and after erosion are listed in Table 8-2. As seen, the binding energy of fluorine, carbon and oxygen of the worn surfaces eroded in PAM and in NaOH are increased respectively relative to that of the original surface (Table 8.2). It has been found that the displacement of higher-binding energy region of functional groups containing oxygen on worn surface eroded in NaOH is greater than that in PAM (Fig.8.14, Fig.8.15 and Table.8.2). It would be accounted for the catalyst effect of the NaOH solution during oxidizing process. In addition, the peak areas of functional groups on the worn surfaces eroded in PAM or in NaOH are increased (Fig.8.14 and Fig.8.15) as compared with the original surface (Fig.8.16). All of these indicate that the element of oxygen might react with carbon adjacent to fluorine, and the degree of oxidization of the rubber surface eroded in NaOH is higher than that in PAM. Therefore, it might be considered that the oxygenated degradation is the main surfacial mechanochemical effect for fluororubber eroded in NaOH.

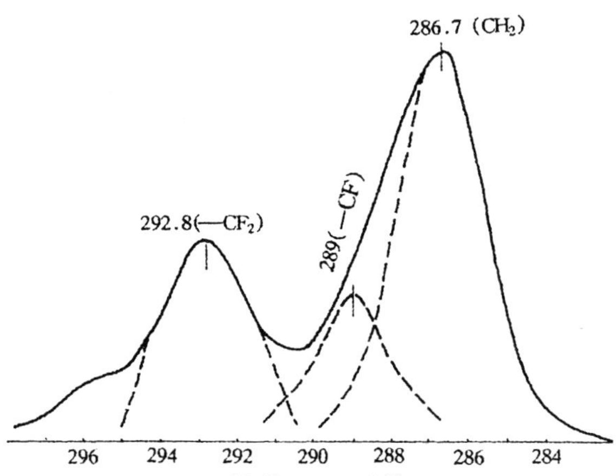

Fig.8.14. XPS spectrum of carbon element of the worn surface of fluororubber eroded in PAM

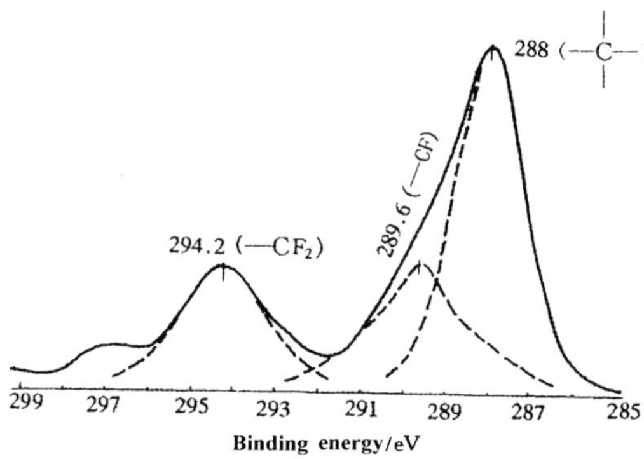

Fig.8.15. XPS spectrum of carbon element of the worn surface of fluororubber eroded in NaOH

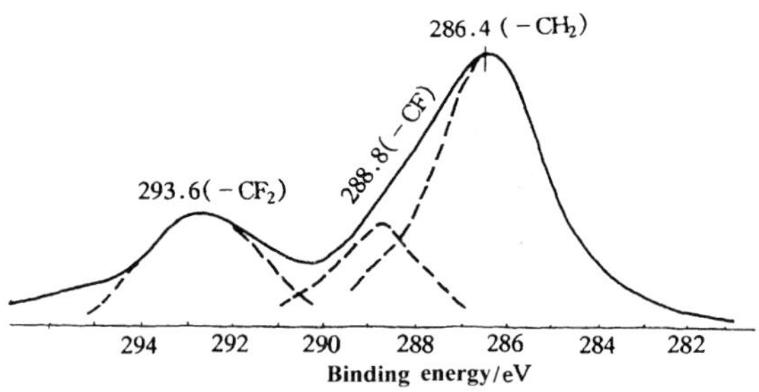

Fig.8.16. XPS spectrum of carbon element of the original surface of fluororubber

Table 8.2. Results of XPS analysis of fluororubber

Specimen surface	Elements composition (wt%)			Binding energy (eV)			
	C	F	O	O_{1S}	N_{1S}	C_{1S}	F(A)
Original surface	35.9	56.8	7.3	533.4	690.0	286.4	602.2
Worn surface (in PAM)	41.6	50.3	8.1	533.8	694.0	286.7	602.0
Worn surface (in NaOH)	44.3	46.4	9.3	534.4	691.6	288.0	604.0

Among the active free-radicals groups resulting from the fracture of macromolecular chains on the fluororubber surface, the product (I) is more prone to be reacted with oxygen in the absorbed water layer. The process of oxygenated degradation of fluororubber is given as follows:

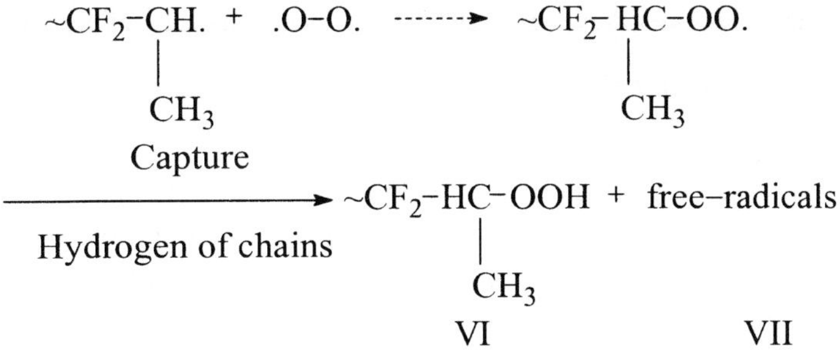

Product (VI) with macromolecular chains containing peroxide groups is easy to be isomerized, and then the oxidants containing carbonyl groups (-C=O) are produced. In addition, product VI might also generate chain termination.

As seen from the above, the oxidizing process of fluororubber is similar to that of natural rubber and nitrile rubber. However, due to the stronger covalent bond of carbon-fluorine, the degree of oxidation of fluororubber is the weakest relative to the above two kinds of rubbers. Based on the analysis of the change of percents of both carbon and oxygen content as well as the spectra peaks of carbonyl groups (-C=O), the oxygenated degradation is not the main surfacial mechanochemical effects for the worn surfaces of fluororubber eroded in H_2O or in PAM, but that for the worn surface eroded in NaOH is not the case.

8.5 Surfacial Mechanochemical Effects of Polyurethane [118]

The structure of macromolecular chains and crosslinking structure of polyurethane are given as follows:

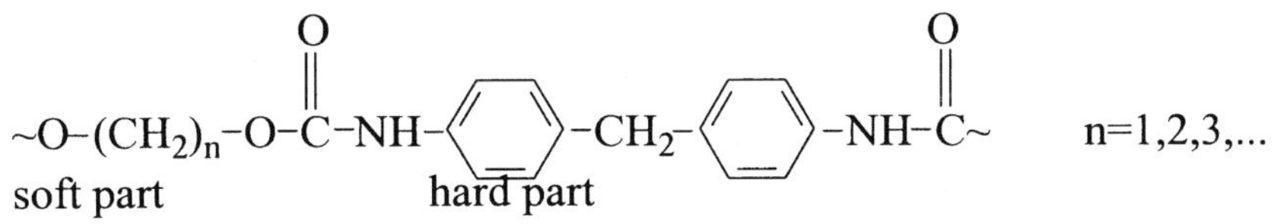

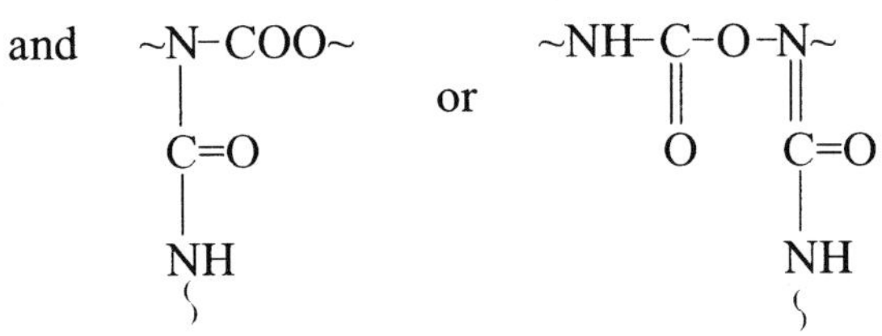

As shown, the macromolecular chains of polyurethane consist of the soft part of polyether and the hard part of polyisocyanaester. It contains cross-linking of isocyanate groups and carbamide groups, and also has hydrogen bonds.

The properties of polyurethane, such as wearability, tearing strength, elasticity, oil-resistance and aging-resistance, are much better, but the character of heat-resistance is poor.

8.5.1. Fracture of Macromolecular Chains

FTIR analysis revealed that the methylene groups (-CH_2-) and ether groups (-C-O-C-) decreased dramatically and the benzene increased (Fig.8.17 and Fig.8.18). It shows that the fracture of macromolecular chains occurred and the fractured site is the covalent bonds of –CH_2-CH_2- at the soft part of the macromolecular chains as well as -C-N- at the connection of hard part and the soft part for the molecule chain.

The position of fracture is given as follows:

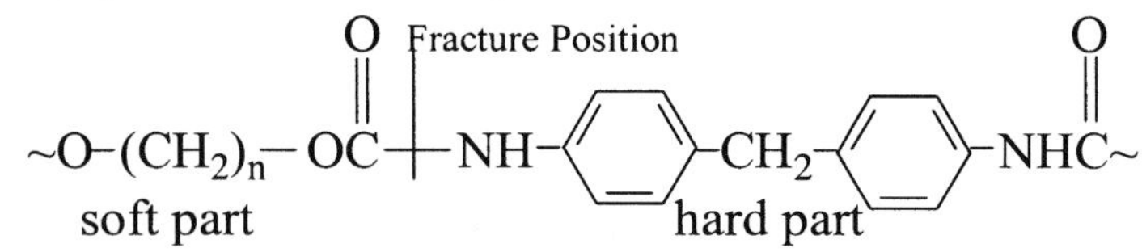

n=1,2,3,...

The free-radicals are:

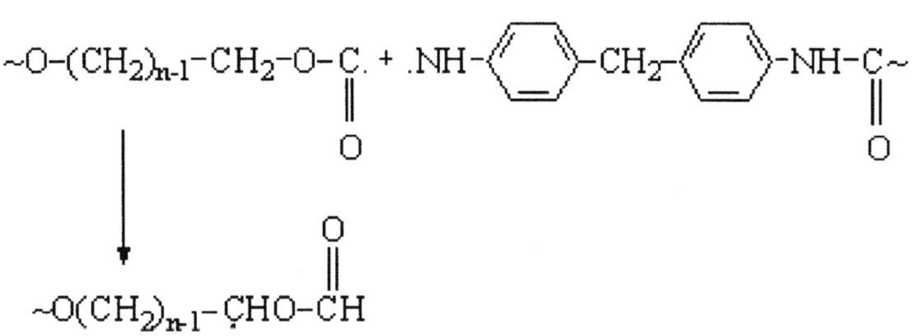

The above conclusions could be proved from the changes of atomic concentration (wt%) of elements by XPS analysis, i.e., the relative weight concentration of carbon is decreased on the worn surfaces eroded in the three mediums respectively and that of nitrogen increased on the worn surface eroded in H_2O (Table 8.3). While the macromolecular chains rupture, free-radicals are produced. Then the ruptured free-radical in the subsurface transfers to the adjacent macromolecular chains and results in the degree of cross-linking being increased, thus, the polyurethane sample might become brittle and produce microcracks. This is one of the reasons inducing fatigue delamination of the subsurface of polyurethane eroded in PAM or in H_2O.

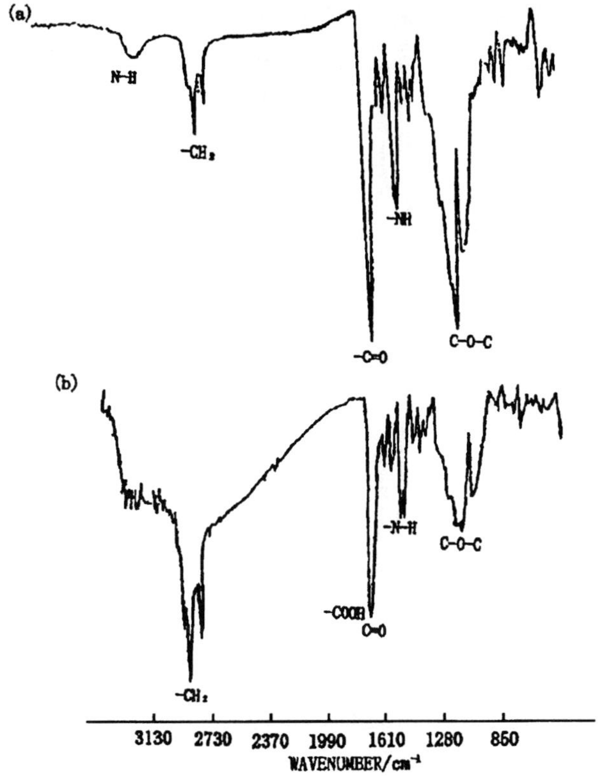

Fig.8.17. FTIR spectra of polyurethane
(a) Original surface; (b) Worn surface (in H_2O)

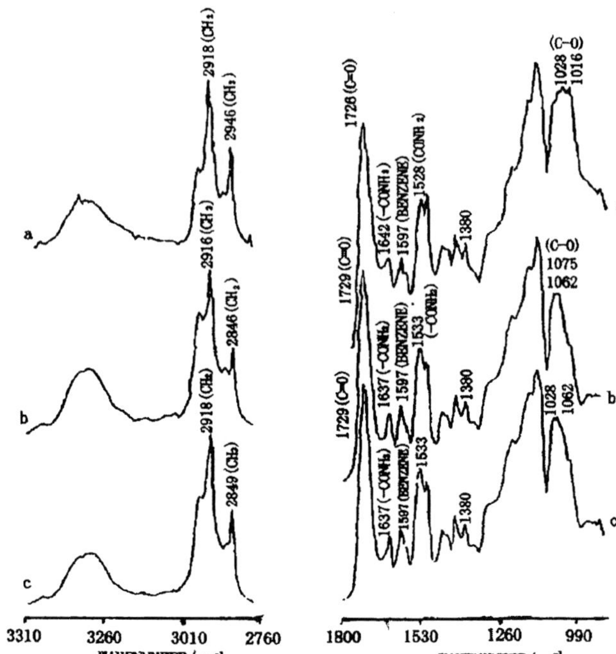

Fig.8.18. FTIR spectra of polyurethane
(a) Original surface; (b) Worn surface (in PAM); (c) Worn surface (in NaOH).

Table8.3. The results of XPS analysis of polyurethane

Specimen surface	Elements composition(wt%)						Binding energy(eV)		
	O	N	Cl	Si	C	Na	O_{1s}	N_{1s}	C_{1s}
Original surface	11.1	2.3	0.4	1.0	85.1	/	533.2	400.5	285.0
Worn surface(in H_2O)	19.3	3.9	0.9	2.4	73.6	/	533.3	400.6	285.0
Worn surface(in PAM)	16.6	1.9	0.4	2.1	79.1	/	532.6	400.3	285.0
Worn surface(in NaOH)	21.0	2.0	0.5	1.4	72.1	3.1	532.3	399.5	285.0

Referring to the micro-rupture models for polymers, i.e. the weakening of the hydrogen bonds and intermolecular Van der Waals forces, as well as the fracture of chemical bonds [155], the fracture process of macromolecular chains could be described further. By the action of microcutting or microtearing of flowing abrasive particles, both the hydrogen bonds and intermolecular Van der Waals forces of the irregularly oriented macromolecular chains were weakened, which cause the slipping of intermolecular chains and the weakening of partial intermolecular attraction. As a result, the degree of physical crosslinking decreased and some chemical bonds in the higher stress-intensity zone were ruptured. It is likely that the fractured site of the chains is at the soft part adjoining the connection of the hard part and the soft part of the macromolecular chains as the most easily cleaved chemical bonds are Carbon-Nitrogen bonds (C-N) (Table 8.4). These covalent bonds have lowest chemical energy and larger bond length.

Table8.4. The bond energy and bond length of covalent bonds

Kinds of bond	Bond energy, kJ/mol	Bond length, nm
C-O	361.2	0.143
C-N	306.6	0.147
C-C	348.6	0.154
C-H	415.8	0.110
N-H	380.6	0.103

The above conclusions could be proved from the decrease of $-CH_2$, C-O-C and the increase in benzene as well as the decrease in weight concentration (wt%) of carbon on the worn surface eroded in the three mediums respectively.

8.5.2 Thermal degradation

Thermal degradation is the thermal decomposition of group, which is resulted from the decomposition of both the allophanate groups and the biuret groups.

Under the acting of impacting, scratching, rolling and rubbing of particles on the surface of polyurethane, the hysteresis set of polyurethane occurs and the maximum shear stress generates at a certain depth beneath the surface [156] when a tangential force (frictional force) acts on the surface of polyurethane. The above phenomena induce the hysteresis loss (heat) and raise the temperature of subsurface which is higher than that of the surface as the heat transfer condition of the surface is much better [114].

As the accumulated heat is generated at a certain depth beneath the surface of polyurethane being in rubbery state, the kinetic energy of the macromolecule could be enhanced. Thus the crosslinked covalent bonds and the molecular chain covalent bonds such as C-N and C-C might be fractured when the kinetic energy of the macromolecular chains becomes larger than the bonding energy of the crosslinked covalent bonds. Therefore, the degree of cross-linking decreased. For example, as the decomposition temperature of allophanate and biuret are 146℃ and 144℃ respectively, these two kinds of groups are decomposed.

Therefore, the thermal decomposition of functional groups in hard part of the macromolecular chains and in crosslinking part of polyurethane occurs. Thermal decomposition might be the main cause for the occurrence of a number of cracks and delaminations on the surface eroded in H_2O or in PAM solution (Figs.6.3 and 6.4).

In addition, by comparison with the polyurethane eroded in water, both the allophanate groups and the biuret groups are easier to be decomposed for polyurethane eroded in PAM as the microcutting is strengthened by the action of cohering particles of polyacrylamide solution. However, since the phenomena of thermal decomposition are very complicated, further work is still needed.

8.5.3 Hydrolysis

Increase in the relative weight concentration for oxygen on the worn surface (Table 8.3) means that oxygenolysis or hydrolysis might have occurred.

From analyzing the polyurethane surface eroded in H_2O by FTIR, it has been found that the number of carbonyl groups on worn surface is decreased and the single pike is changed into double pikes. Moreover, the carboxyl groups can be found at the position of these pikes (Fig.8.17). The above proves the occurrence of hydrolysis during the wear process of polyurethane eroded in H_2O as hydrolysis is one of the causes for the appearance of carboxyl groups.

As for the worn surface of polyurethane eroded in PAM and in NaOH respectively, it has been found by FTIR analysis (Fig.8.18) that the amide groups and carbonyl groups are increased, which means the emergence of hydrolysis. Moreover, the degree of hydrolysis of the surface eroded in NaOH is the highest by comparison with that in the other two mediums.

The strongly polarized carbamate groups (-NHCOO-) in macromolecular chains could result in hydrolytic degradation once the temperature is high enough [154]. The experimental temperature of mediums is about 63℃-67℃ in slurry tank, which is lower than that in annular pipe. Moreover, the mechanical stimulation of the flowing particles on the surface eroded in mediums might strengthen the interaction between the macromolecular chains on the surface and the mediums, and then results in the hydrolysis of carbamate groups.

The process of reaction is given as follows:

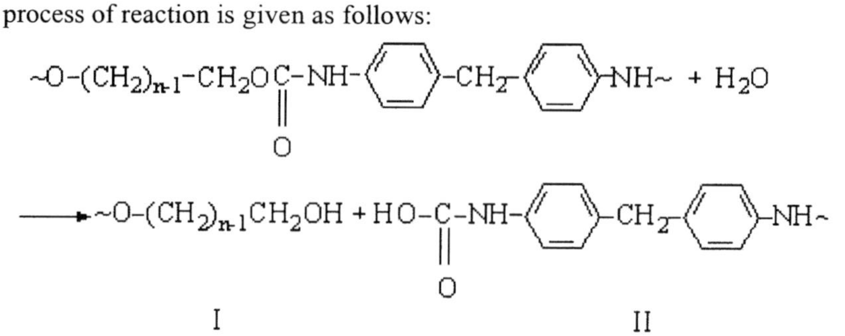

I II

The reaction product (I) is easy to be flushed away as it is located in the soft part of chains. However, the reaction product (II) is situated at the hard part of chains.

Due to the fluid medium of PAM being exhibited acidic (pH=6), the number of carbonyl groups and amide groups on the worn surface eroded in this medium is increased (Figs.8.18 and Fig.8.19). Therefore, the degree of hydrolysis in PAM is higher than that in H_2O. As the hydronium ion (H^+) in acidic solution is combined at the polar carbonyl oxygen and made positively charged ($C=OH^+$), the electronic cloud is transferred to the oxygen of the carbonyl group and made the carbonyl carbon atom strongly positively charged. Thus, it is easy to produce nucleophilic addition reaction. Moreover, the action of cohering particles of solution (PAM) could make the interaction between surface of polyurethane and flowing particles much stronger. Therefore, a lot of reaction products (I) are flushed away. However, the reaction products (II) still exist on the surface as they are located in the hard part of the macromolecular chains of polyurethane, which induce the number of carbonyl groups and amide groups to be increased (Fig.8.18). Because the molecular weight of the reaction products is decreased, the mechanical properties of polyurethane becomes so poor that the wear resistance is decreased

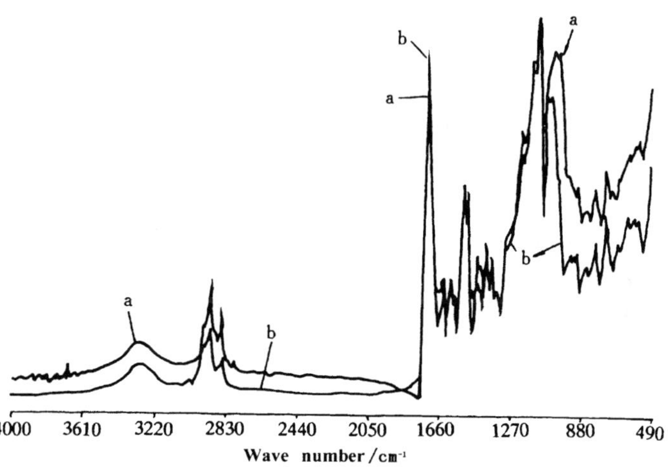

Fig.8.19. Comparison of FTIR spectra of the worn surface eroded in PAM with the original surface of polyurethane

(a) Original surface; (b) Worn surface

The reaction process of hydrolysis of polyurethane eroded in PAM is given as follows:

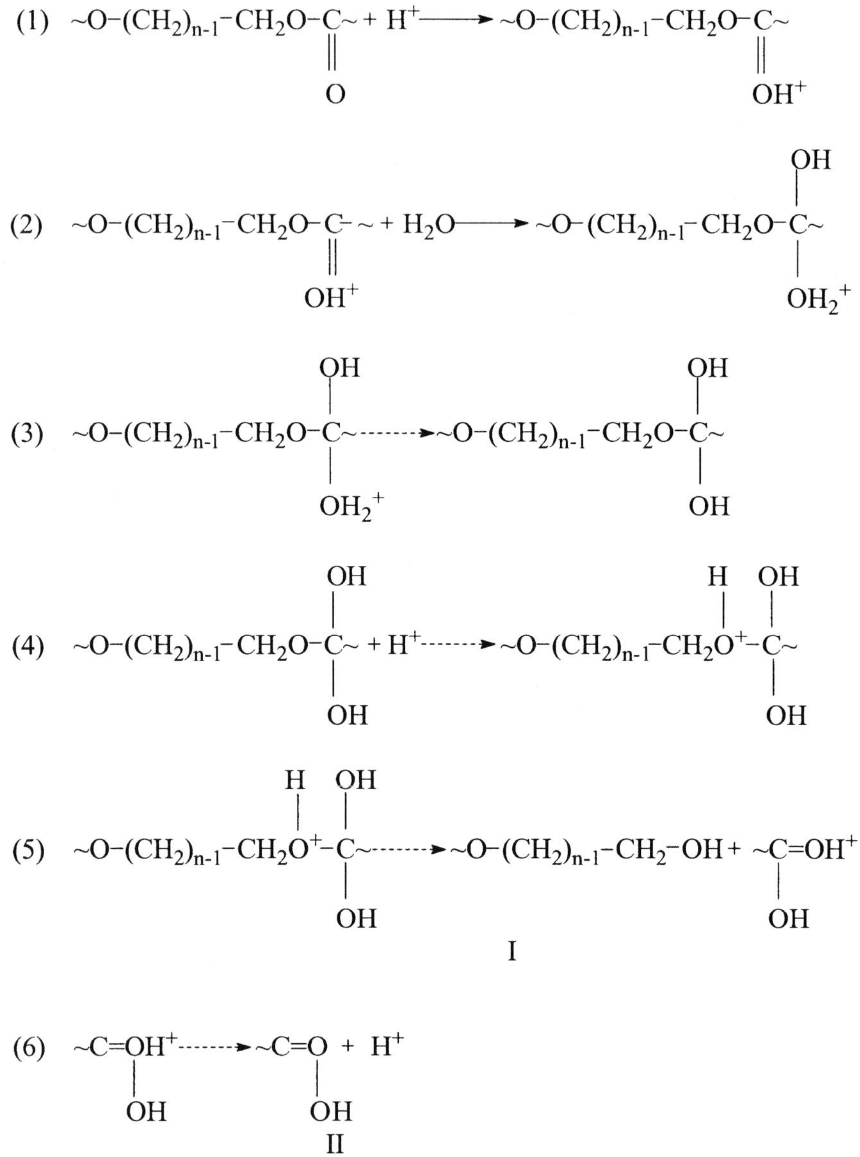

The whole process could be expressed briefly as follows:

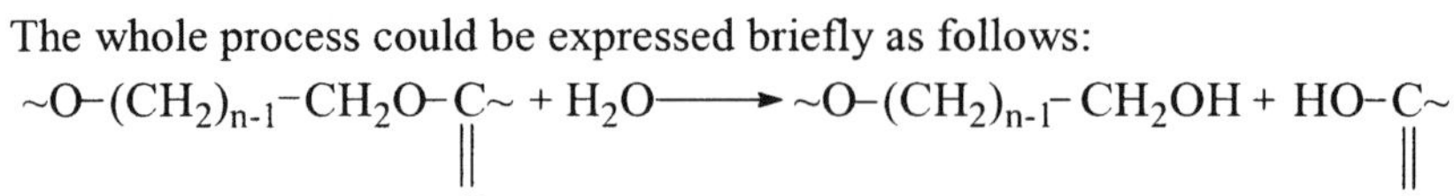

I II

Among the three different mediums, due to the medium of NaOH being exhibited basic (pH=12) and containing sodium, the hydrolysis for the surface eroded in NaOH is most completely. Moreover, the groups of -OR' in RCOOR' are easily replaced by the high-nucleophilic and basic groups of -OH, which induces carboxyl groups (COOH) to be produced. As soon as the occurrence of carboxyl groups, the reaction between carboxyl groups and basic group of -OH occurs and then produces salt. All of these reaction processes could be proved from the existence of sodium as well as the increase in the number of carbonyl groups and the amide groups on the worn surface eroded in NaOH (Table 8.3, Fig.8.18 and Fig.8.20). For the above reasons, the increase in number of carbonyl groups and the wear rate is the most, moreover, no crack and indentation occurs on the worn surface eroded under this condition.

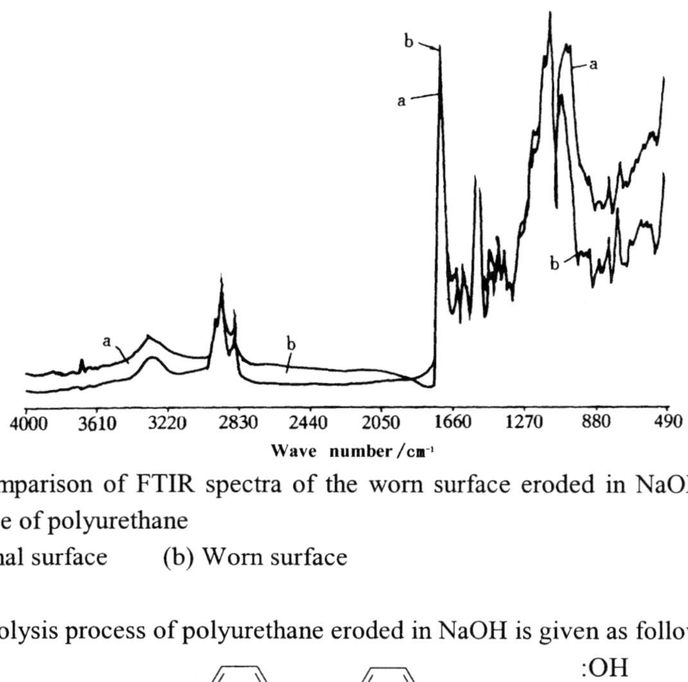

Fig. 8.20. Comparison of FTIR spectra of the worn surface eroded in NaOH with the original surface of polyurethane

(a) Original surface (b) Worn surface

The hydrolysis process of polyurethane eroded in NaOH is given as follows:

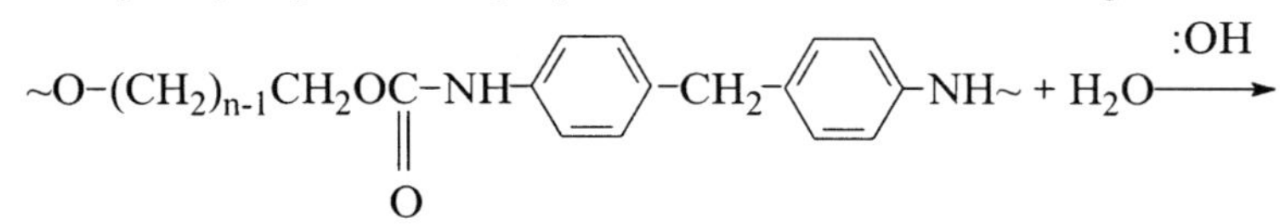

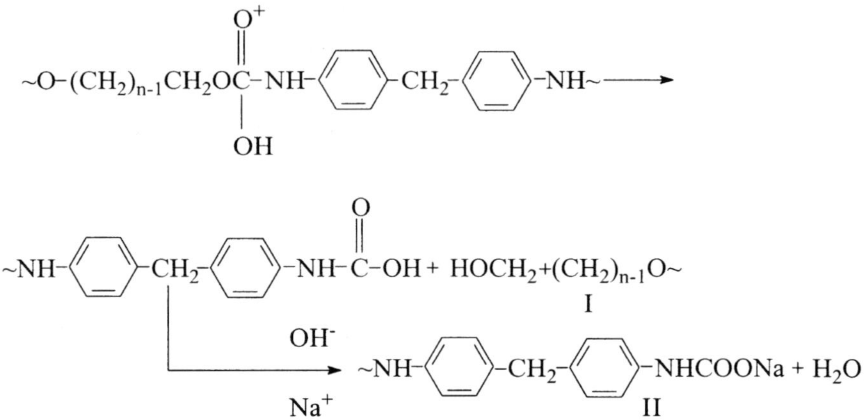

In summary, the degree of hydrolysis of polyurethane in PAM and in NaOH respectively is higher than that in water. This conclusion could be proved by the fact that the increment of carbonyl groups of polyurethane in PAM or in NaOH solution, and the decrement of -C=O of polyurethane in water as observed in the FTIR spectra as shown previously.

8.5.4 Surfacial Oxygenated Degradation

As seen from the peaks of C_{1s} levels (Fig.8.21, 8.22(a) and 8.23(a)), the sorts of carbon-oxygen functional groups, including, $-O-C(=O)-O-$, $R-C(OH)=O$, C-O, C=O on the worn surfaces eroded in the three different mediums respectively are more than that on the corresponding original surface (Fig.8.24). Moreover, the peak area in the region of high-binding energy of C_{1S} levels on worn surface eroded in PAM and in H_2O respectively is much larger than that on the original surface (Fig.8.21, Fig.8.22 (a) and Fig.8.24). It shows the surface being degraded by oxidation.

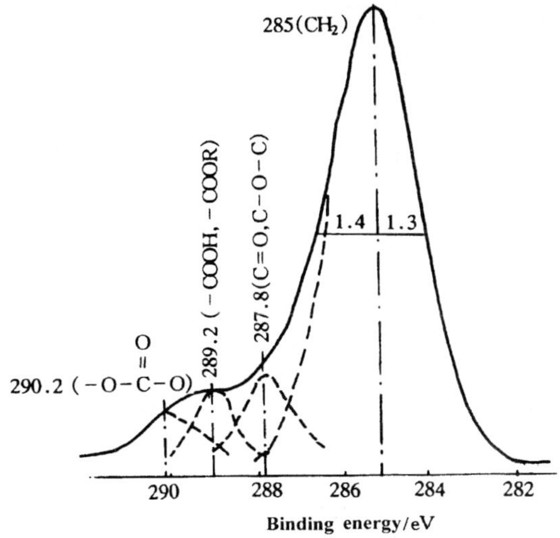

Fig.8.21. XPS spectrum of C_{1S} on the worn surface of polyurethane eroded in H_2O

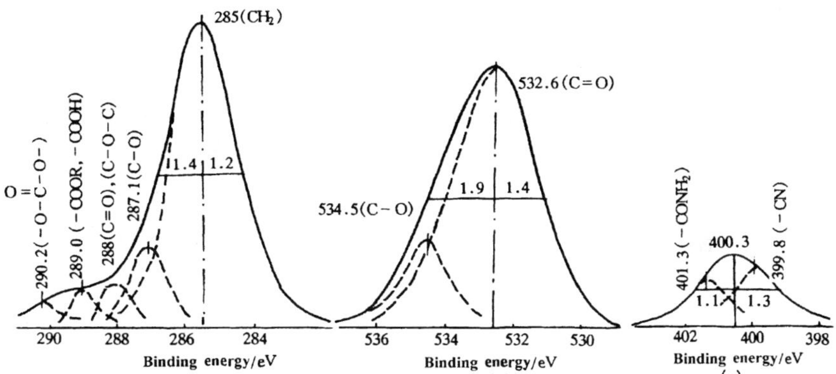

Fig.8.22. XPS spectra of worn surface of polyurethane eroded in PAM
(a) Carbon element (C_{1S}); (b) Oxygen element (O_{1S}); (c) Nitrogen element (N_{1S})

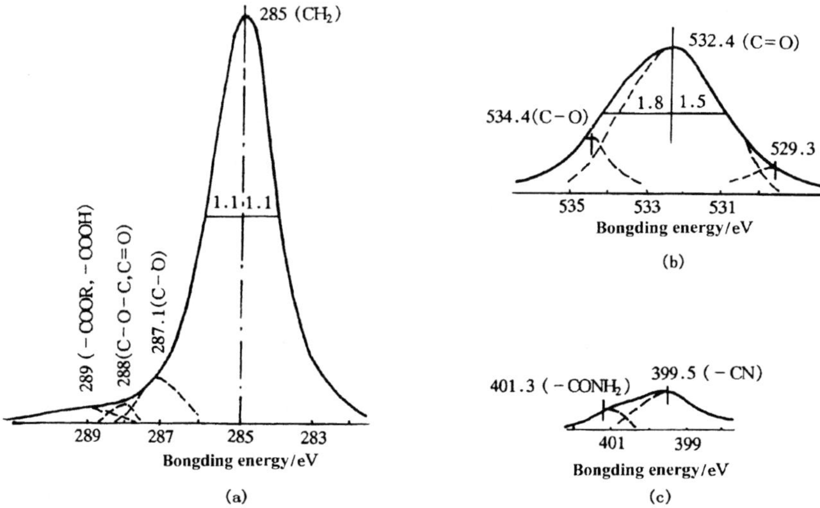

Fig.8.23. XPS spectra of worn surface of polyurethane eroded in NaOH
(a) Carbon element (C_{1S}); (b) Oxygen element (O_{1S}); (c) Nitrogen element (N_{1S})

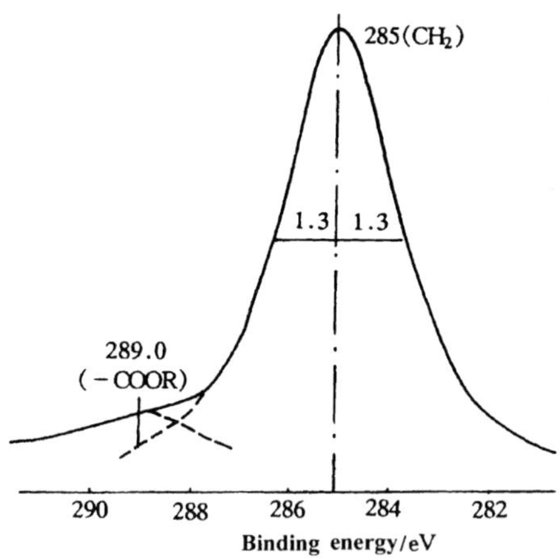

Fig.8.24. XPS spectrum of C_{1S} on the original surface of polyurethane

In addition, from the infrared spectra, it can be found that the methylene groups decrease and the "V" shapes occurs at higher-wave-numbers than for the worn surface eroded in water (Fig.8.17). It shows that some of methylene groups (-CH_2-) at the soft part of the macromolecular chains are oxygenated and the oxidative products are carboxyl groups being bonded each other by hydrogen bonds [154].

Due to the rupture of the covalent bonds of -CH_2-CH_2- at the soft part as well as -C-N- at the connection of the hard part and the soft part of macromolecule chains, the active free-radicals emerge. The ruptured active free-radicals are apt to react with oxygen in water. The main oxidative product is peroxide. However, by means of a series of reactions and transmutations, the peroxide might turn into oxidative product containing carbonyl groups (-C=O). Because the molecular weight of oxidative products is decreased, the mechanical properties of polyurethane become so poor that its wear resistance is decreased; some of the oxidative products containing carbonyl groups are flushed away.

The free-radicals reaction process of oxygenation is given as follows:

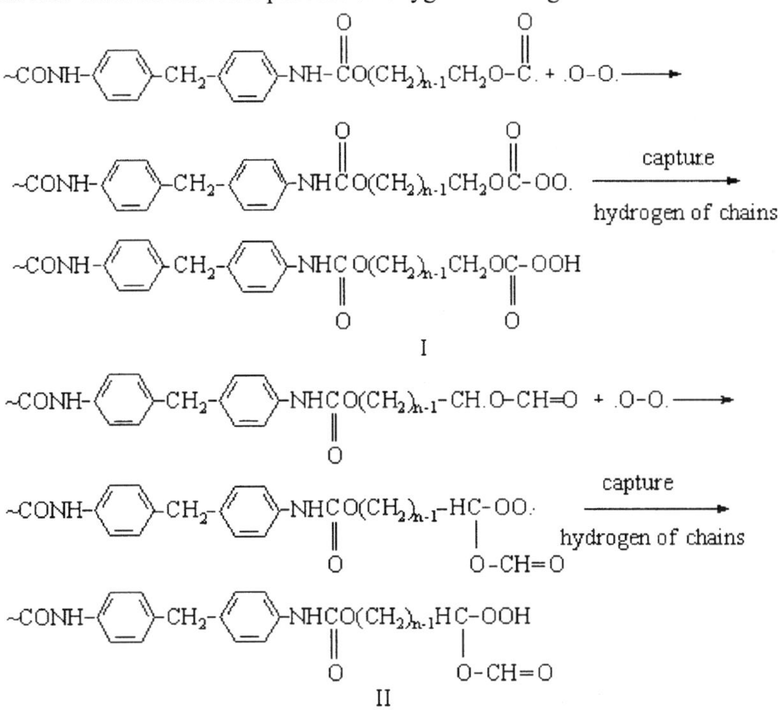

The reaction products I and II are easily flushed away as they locate at the soft part of the macromolecular chains. Therefore, the carbonyl groups (-C=O) of ester groups (-COOR) on the worn surface eroded in H_2O are decreased in large numbers (Fig.8.17). On the contrary, the carbonyl groups on the worn surfaces eroded in PAM and in NaOH respectively are increased (Fig.8.18). Therefore, referring to the above free radicals reaction processes, it could be inferred that the degree of oxygenation for the surface of eroded in H_2O is the most by comparison with that in the other two fluids according to the change of carbonyl groups. The above results could prove that the degree of hydrolysis of polyurethane eroded in PAM or in NaOH solution is higher than that eroded in H_2O, which causes a lot of methylene groups ($-CH_2$) at the soft part of macromolecular chains to be flushed away. Therefore, the oxygenation of active free-radicals for the surfaces eroded in of PAM or in NaOH solution is constrained. So, it would be concluded that the oxygenated degradation is not the main surfacial mechanochemical effect of polyurethane samples eroded in PAM and in NaOH.

During the abrasive erosion, under the action of microcutting and impacting of the particles, the ether groups (-C-O-C-) being in excited state are apt to react with oxygen absorbed on the water layer. The more likely oxidizing position is on the α-carbon hydrogen bonds, that is, $-CH_2-CH-O-$. The oxidizing products are further isomerized,

α |

HOO

O

|

and the final products containing -O-C-O- are produced.

The oxidizing process is given as follows:

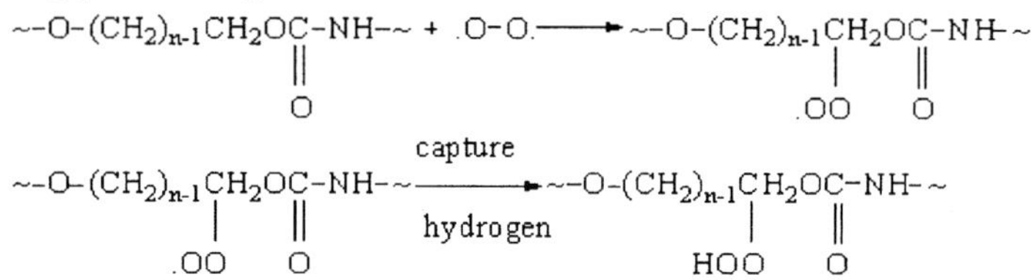

Chapter 9

WEAR OF METAL BY RUBBER

9.1 Literature Survey

It is well known that the wear value of a soft solid surface is usually larger than that of its hard counter face in the friction couples. However, the wear of metal by rubber, an interesting and unexpected wear phenomenon, was observed for some rubber-metal friction pairs. Unfortunately, Even though this special wear phenomenon was discovered as early as the 1960's [87], it escaped the attention so far. Research work in this aspect is quite insufficient.

As the rubber-metal frictional couples are used widely in a variety of machines and the metal components among them are usually more expensive than the rubber counter parts, study in the wear of metals by rubber is of vital importance. Therefore, it is necessary to understand the development of this subject in the past decades including our work carried out of recent years.

9.1.1 Wear Behaviors of Metal by Rubbers

King and Lancaster [157] have investigated the wear of metals by elastomers using a modified pin-on-disc apparatus in the presence of a clean fluid or a dispersion of abrasive particle. Experiments were conducted with a metal ball sliding against a disc of elastomers. The various elastomers used are listed in Table9.1 and some thermoplastic polymers were also included for comparison purpose.

It has been found that the wear of metal by elastomers is dependent on the state of particles on the counter face (free abrasive or embedded abrasive), and the Shore hardness, elastic modulus, or resilience of the elastomers. The wear rate of steel by elastomers in water without abrasive was less than 10^{-7} mm^{-3} N^{-1} m^{-1}. When abrasive was introduced, the wear rate of steel plotted against Shore hardness of various elastomers is approximated to a power relation, which increased with the increase in elastic modulus of the elastomers up to about 20 MPa (Fig.9.1 and Fig.9.2). However, the relation between the wear rates of steel by elastomers and elastomer resilience is reversed and just significant at around the 0.1% level (Fig.9.3) [157]. In Figs.1, 2 and 3, the members refer to the individual polymers is listed in Table 9.1

Table 9.1 Experimental materials [157]

No.	Type		Hardness (shore A)	Resilience (%)	Elastic modulus (MPa)
	Elastomers				
1	Natural rubber		38	48	1.5
2	Epichlorhydrin(Herclar)		77	4	7.5
3	Styrene-butadiene (SBR)		73	36	6
4	Polyurethane		85	4	12.5
5	Polychloroprene (Neoprene)		68	13	4.9
6	Polyacrylate (Krynac882 X 2)		82	4	10
7	Polysulphide (Thiokol ST)		75	28	7
8	Chlorosulphonated polyethylene(Hypalon)		84	6	12
9	Fluorocarbon (Viton B)		75	4	6.7
10	(different degrees of cure)		75	4	7
11			82	6	10
12	Silicone	(Silastic 35)	42	41	1.7
13	(different filler contents)		55	41	3
14	(Silastic 55)		56	36	3
15	(Silastic 75)		60	38	3.5
16	Nitrile (different filler		62	7	3.8
17	contents and various ratios	(Hycar 1002)	68	16	5
18	of acrylonitrile to butadiene)	(Krynac 802)	71	17	5.5
19	(Hycar 1041)		76	5	7
Thermoplastics					
A	Polytetrafluoroethylene				490
B	Polypropylene				1120
C	Acrylonitrile-butadiene-styrene				1400
D	Acetal				2870
E	Polyethylene (LD)				250
F	Thermopiastic polyurethane				470
G	Plasticized polyvinyl chloride				15

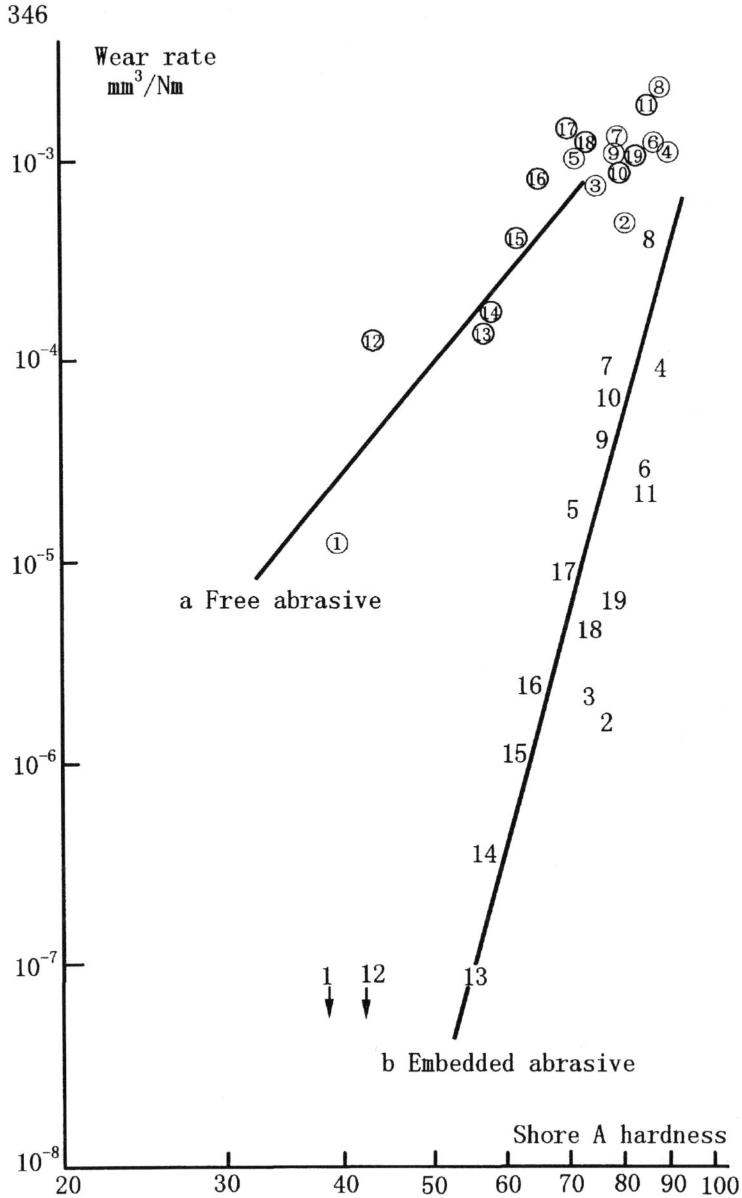

Fig.9.1 Variation of metal wear with elastomer hardness:
Curve (a), steel, free abrasive, wear rate proportional to $H^{5.2}$, correlation coefficient 0.90; Curve (b), bronze, embedded abrasive, wear rate proportional to $H^{10.2}$, correlation coefficient 0.87 [157].

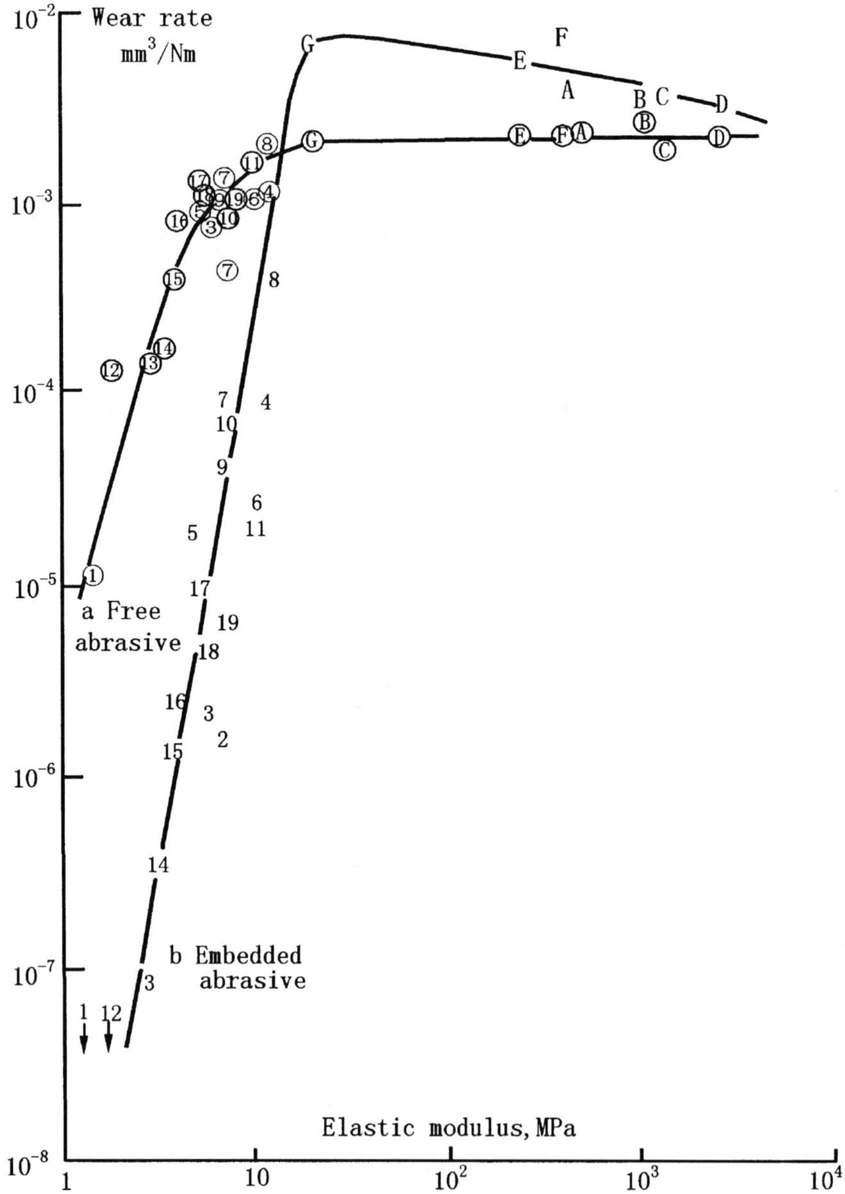

Fig.9.2 Variation of metal wear with elastic modulus for elastomers and polymers: Curve (a), steel, free abrasive; Curve (b), bronze, embedded abrasive [157].

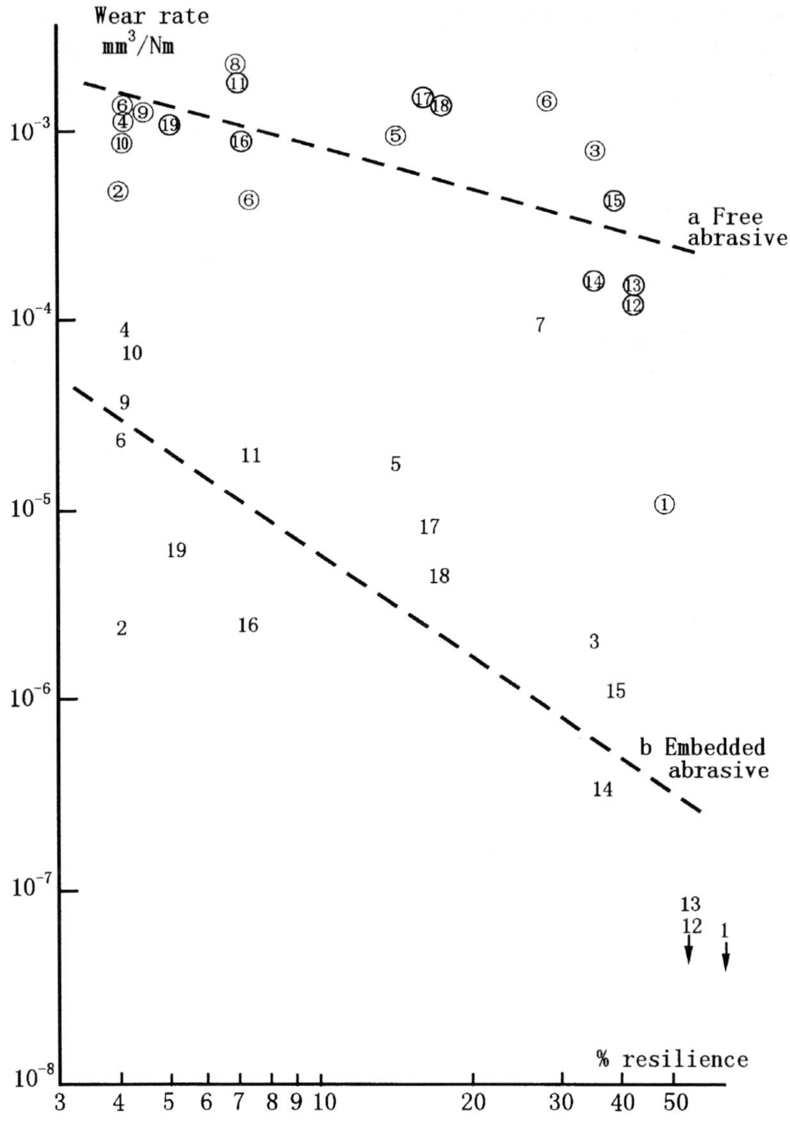

Fig.9.3 Variation of metal wear with elastomer resilience:
curve (a), steel, free abrasive, wear rate proportional to $R^{-0.8}$, correlation coefficient -0.62; curve (b), bronze, embedded abrasive, wear rate proportional to $R^{-1.85}$, correlation coefficient -0.70 [157].

Based on the results above, King et al. [157] considered that the Shore hardness and the elastic modulus of elastomers have importance influence on the rate of wear of steel by elastomers. Furthermore, they proposed three decisive factors: (1) the total amount of the abrasives embedded in elastomer; (2) the penetration depth of the abrasive particles into the elastomers under load; (3) the strength of adhesion of the embedded particles to the substrate.

Using various cylindrical alloy-steel indenters to puncture the surface of rubber blocks repeatedly, AB-Malek and Stevenson [158] examined the wear of alley-steel by rubber and found that the dominant factors influencing the wear rate of metal by rubber include the hardness of both metal and rubber, carbon concentration and crosslinking level of the rubber; the self-lubricating ability of rubbery boundary layer and the stability of metal oxide layer. Charrier et al. [159] observed that a metal needle could be worn by elastomer, if the needle penetrates the rubber repeatedly. Of recent years, Zhang and coworkers [160,161] have investigated the wear of two kinds of steel by three kinds of rubbers in different media. The details are presented in the subsequent Section 9.2.

9.1.2 Wear of Mechanisms of Metal by Rubbers

King and Lancaster [157] considered the mechanism of wet abrasive wear of steel ball by rubber to be that the hard particles embedded in rubber surface scratches the metal surface during sliding process. However, AB-Malek and Stevenson [158] proposed that the wear of steel by rubber is resulted from the fatigue of metal oxide under high stress during repeated punctures. Moreover, they observed a lubricating layer of rubbery materials being formed on the metal surface. In order to clarify this phenomenon, experiments were carried out by using indenters made from titanium alloy, pure titanium (99% Ti), tool steel and tungsten carbide respectively to puncture different vulcanizate rubbers. It has been found that the molecular segments of the freshly ruptured rubber were adhered to the metal surface under the action of van der Waals' secondary intermolecular forces, and then a lubricating adhesion layer of rubbery materials was formed on the indenter surface. The free radicals of segments in the lubricating layer reacted with the metal oxide surface and produced a metal oxide-polymer complex, which was weaker than the metal oxide surface itself and can be detached more easily from the surface. A general trend is towards increasing the wear rate with decreasing the hardness of metal, because a hard metal means in fact a high local plastic yield stress. Therefore, the stable oxide layers of metal react much less readily with rubber radicals and a relatively thick and continuous layer of rubbery material acts as a lubricant and retards wearing. But there is an anomaly in that the softest metal (titanium) shows much lower wear rates than expected for its hardness because it is noted for the stability of its oxide layer.

Charrier et al [159] found that the transfer layer of rubber on the needle surface reduced the wear rate of needle.

Gent and Pulford [94] investigated the wear mechanism of steel by cis-polyisoprene

(IR) and cis-polybutadiene (BR) respectively using a blade abrader designed by Thomas [65]. They observed that the wear of steel blade by a rubber wheel took place much more rapidly on a cis-polyisoprene surface than that on a cis-polybutadiene surface, and more seriously in inert atmosphere than in air. The long-lived radicals formed in IR appear to react readily with steel, presumably forming an iron-carbon compound, which is removed along with the rubber debris. In contrast, the more reactive radicals formed in BR appear to undergo mainly reaction within the rubber so that steel suffers much less wear in this case. As for the peroxy radicals, its reactivity is apparently lower for BR than for IR so that attack on steel is less for material in air.

In order to examine the process of mechanochemical reaction in detail, Gent and Pulford [95] conducted further experiments by using three kinds of metal razor blades (stainless steel, nonferrous alloy and bronze) against the surfaces of six kinds of rubbers (SBR, standard Malasian rubber (SMR) , isobutene-isoprene rubber (IIR), trans-1,5-polypentenamer rubber (TPR), polybutadiene (BR) and ethylene-propylene (EPR)), respectively. It has been found that wear rate of steel by rubber can vary by a factor of 50 or so when both the hardness of the rubber and the frictional force are kept constant (Table 9.2). Moreover, in a nitrogen atmosphere, the wear rate of a metal scraper is generally increased by a large factor, between 5 and 50 times, depending on the rubber against which the scraper slides. It is attributed mainly to the greater stability of carbon radicals, in general, in comparison with corresponding peroxy radicals. However, an apparent exception is the rate of wear against IIR compound, it was found to be greater in air than in nitrogen, by a factor of about 3. This anomalous behavior is ascribed to enhanced stability of the peroxy radicals in IIR compared to the carbon radicals formed by molec1liar rupture.

Gent and coworkers [95] pointed out that the wear rate of metal is closely related with the stability of polymeric radicals, which cause metal wear by a combination of chemical reaction and detachment of metallic fragments. When the radicals are highly reactive, it is thought to take pad primarily in internal polymer reactions and thus cause relatively little wear of the metal scraper. On the other hand, relatively stable polymer radicals appear to attach metals vigorously. They also found that the rate of wear of metal increased markedly as the hardness of the rubber compound increased by incorporating more carbon black, but the mechanism has not yet been identified.

Table 9.2 Wear rate of metal razor blades by six kinds of rubbers [95]

Type of rubbers	hardness（Shore A）	Test atmosphere	Wear rate（$X10^{-16}m^3/r$）
SBR	75	Air	13
SBR	75	N_2	100
SMR	60	Air	0.75
SMR	60	N_2	40
IIR	57	Air	17
IIR	57	N_2	5.2
TPR	67	Air	0.70
TPR	67	N_2	5.2
BR	66	Air	0.30
BR	66	N_2	2.5
EPR	56	Air	0.25
EPR	56	N_2	1.3

Some direct evidences of reaction of macromolecular radicals to metal surfaces have been found by Gent and Rodgers [96]. In their experimental study, metal powders (iron, zinc and aluminum) wear incorporated into various elastomeric materials (SBR, NR, BR and EPR) and the mixture were subjected to intense mechanical shearing. Figure 9.4 shows the UV-visible spectra of samples of SBR which have been subjected to intense shearing with and without iron powder being present. By using this control sample as a reference, a new absorbance is found at 340 nm for the samples that were sheared with iron powder (Fig. 9.5). It is noteworthy that some iron-containing organic compounds have absorbances in this region, although these compounds have not been identified more explicitly. From Fig. 9.5, it is shown that the absorbance at 340 nm increases continuously with the extent of sharing, denoted by the number of milling passes for a chemical reaction.

Gent and Rodgers [96] also observed that the amount of iron or zinc taken up by the rubber depends on the kind of macroradical by molecular rupture, namely, relatively long-lived radicals (SBR, NR) are associated with greater metal pick-up compared with more reactive radicals (BR, EPR).

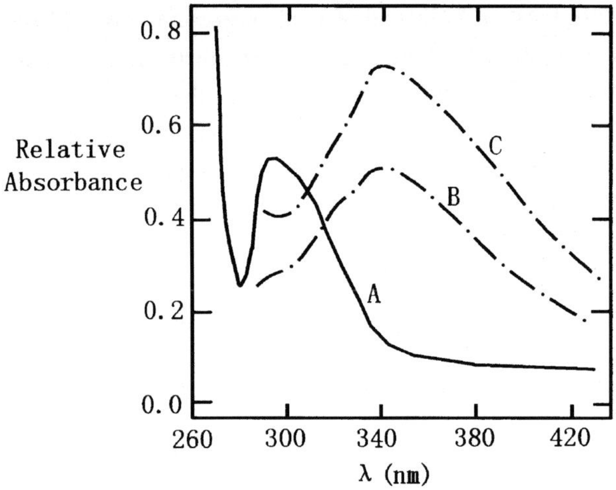

Fig.9.4 UV-visible spectra of styrene-butadiene rubber:
(A) control sample;
(B) sample sheared in air with 100 parts by weight of iron, referenced to the control;
(C) sample sheared in air with 300 parts by weight of iron, referenced to the control [96].

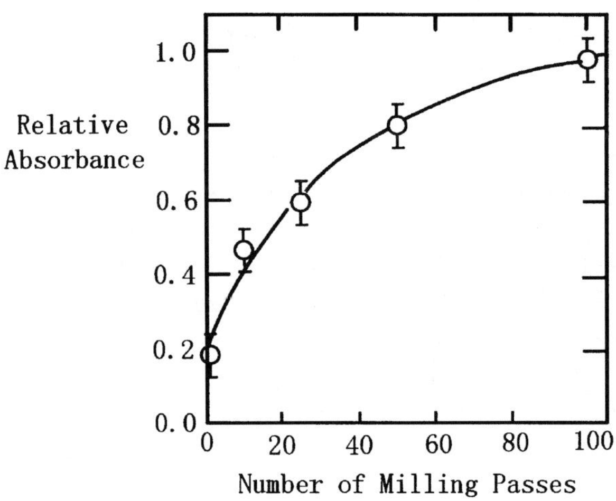

Fig.9.5 Relative absorbance at 340nm versus amount of shear for styrene-butadiene rubber with iron powder (300 phr) [96].

To sum up, the wear process of metals by elastomers is a complex phenomenon, which involves several processes, such as physical, mechanochemical and thermal-chemical and so on. To clarify this phenomenon is of importance to the design and use of the rubber-metal friction assemblies. Unfortunately, it has received too little attention in the scientific and engineering circles in the past decades. Although marked progress is being made in this subject of recent years, there are still some key problems need to be addressed, such as, the processes of reactions among the macroradicals of elastomers, metal atoms and metal oxides; interaction between the physical effects and the chemical effects of the wear of metal by elastomers, etc.. It is expected that these studies would open up the prospects for developing some new techniques of metal processing and surface engineering.

9.2 Wear of Metal by Rubber under Boundary Lubrication Condition

In this section, it is mainly discussed the wear of steel T10 by nitrile rubber (NBR) under boundary lubrication condition with mineral oil [160].

9.2.1 Experimental Methods [160]

Experiments test were carried out using a pin-on-disc sliding wear test machine (Fig.9.6). The pin with a diameter of 5 mm was made of steel T10, the composition of which is given in Table 9.3. The disc was made of nitrile rubber (NBR) with a diameter of 60mm. The mechanical properties of the NBR materials are shown in Table 9.4.

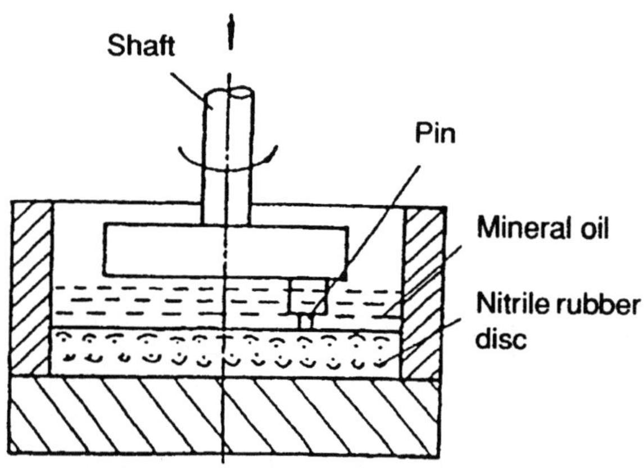

Fig.9.6 Sketch of wear test machine

Table 9.3 Composition of steel T10 (%)

Material	C	Mn	Si	S	P
T10	0.95~1.04	≤0.40	<0.35	≤0.030	≤0.035

Table 9.4. Mechanical properties of NBR materials

Harness (Shore A)	Tensile Strength (MPa)	Elongation at break (%)	Coefficient of resilience (%)	Density ($g.cm^{-3}$)
80	14.71~29.42	400~800	20~50	0.96~1.20

Under the testing conditions of a normal load of 12N and a sliding speed of 80 r/m, experiments were performed using 10ml paraffin mineral oil as lubricant at room temperature (20℃). The bulk temperature of the oil was 50℃ during the experimental period of 35h. The surface topographies of both rubber and metal samples were examined by scanning electron microscope (SEM, Cambridge S-360). The chemical states of the elements in both surfaces were analyzed by X-ray photoelectron spectroscope (XPS, PHI-5300). The sputtering speed of the XPS analysis was 3nm/min and the sputtering point was located at the center of the metal surface. The changes of functional groups in the rubber samples were analyzed by Fouriey transform infrared spectrometer (FT-IR, Perkin Elemer System 2000), diffuse reflectance in air. The IR scan conditions were temperature of 28℃ and relative humidity of 54%Rh. In addition, the mineral oil was analyzed before and after testing by FT-IR (Nicolet SX). After wear testing, this oil was sampled from the bulk fluid by syringe.

9.2.2 Analyses of Samples [160]

9.2.2.1 SEM Examination

Obvious parallel ploughs, scratches, indentations, pits and some films were observed on the worn surface of steel T10 (Fig.9.7(a)). The morphology of the worn surface of the nitrile rubber is shown in figure 9.7(b). Apparently, the nitrile rubber swelled during the wear processes and some lacerations were observed on the worn surface.

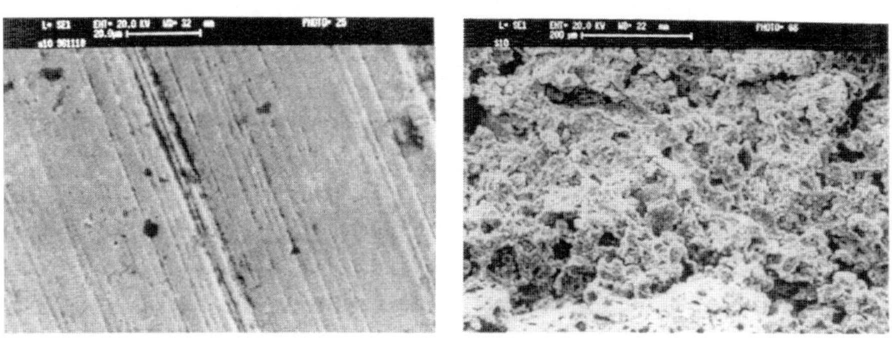

(a) Worn surface of steel T10 (b) Worn surface of nitrile rubber

Fig.9.7 SEM micrographs of worn surfaces (×1000)

9.2.2.2 FT-IR Analysis

(1) Mineral oil

The main constituent of mineral oil is long-chain alkane. Figure 9.8 illustrates the infrared absorption spectrum of the mineral oil before and after the wear tests. The decrease in absorption intensity of CH_2 (2923cm^{-1}, 1463cm^{-1}) and CH_3 (2853cm^{-1}) in the used oil indicates that some of C-C and C-H bonds in the lubricating oil were broken. The occurrence of absorption peak of C=O (1732cm^{-1}) and C-O (1073cm^{-1}, 1121cm^{-1} and 1285cm^{-1}) in the used oil shows that the mineral oil was oxidized during sliding process. This is accounted for by C-C bond in the long-chain molecular of mineral oil being ruptured and producing active free radicals under the action of shear stresses. The free radicals are easily oxidized by the oxygen in air to generate oxides. However, apparent vibration of the C=C bond was not observed, which means that the frictional heating did not induce dehydrogenation of the mineral oil to produce unsaturated double bind. Moreover, any C=C bonds would be rapidly oxidized if they were formed.

worn surface.

The O_{1S} peak is wider for the worn surface than for the original surface (Fig.9.10). This confirms that the worn surface is oxidized during the test.

These results are consistent with those of FT-IR analysis, which shows that there is only C, O and N in the original surface. After the tests, Fe exists on the worn surface in the form of Fe-C, Fe_2O_3, FeOOH and Fe_3O_4 (Fig.9.11), which indicates that Fe and iron oxides react with the macromolecular chains in the worn surface.

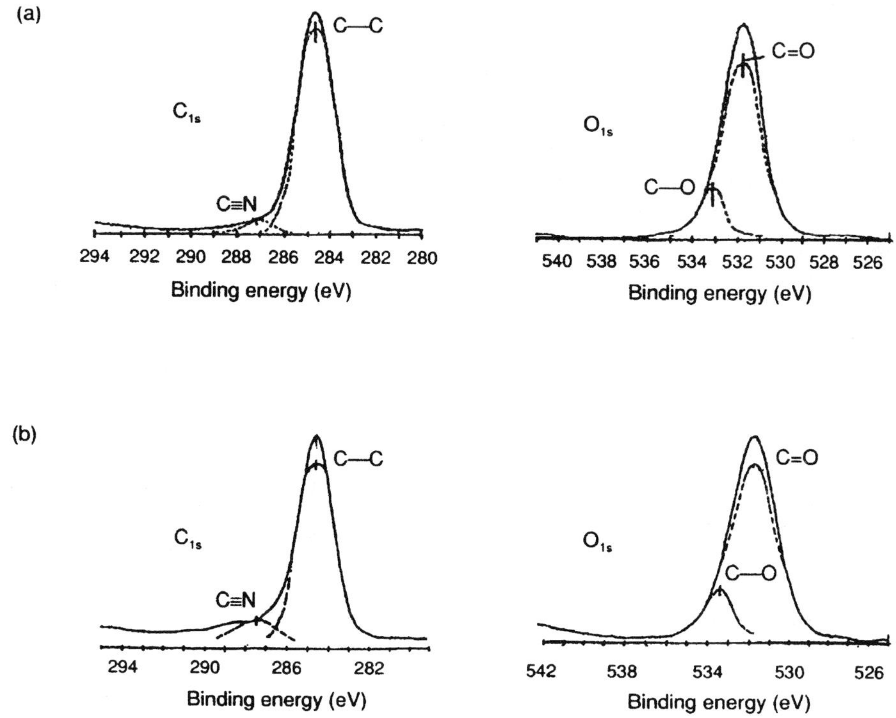

(a) C_{1S} and O_{1S} on the original surface (b) C_{1S} and O_{1S} on worn surface

Fig.9.10. XPS spectra of nitrile rubber

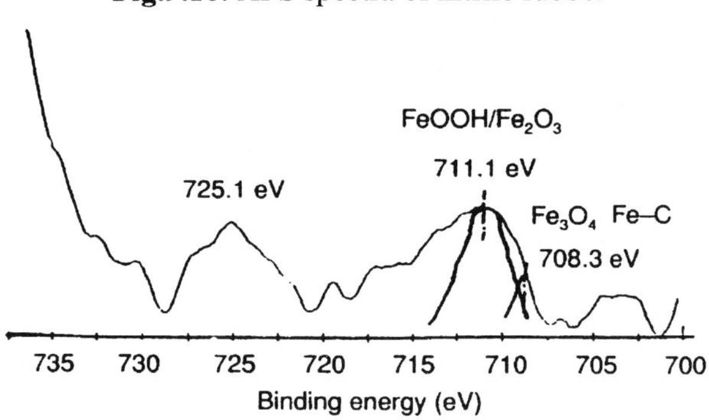

Fig.9.11. $Fe_{2P2/3}$ spectrum on worn surface of nitrile rubber

(2) Surface of Steel T10

The elements in the original and worn surfaces of steel T10 are given in Table 9.5. It can be seen that the atomic concentrations of the three elements have changed. The atomic concentration of C_{1S} increased in the worn surface, indicating that some of the carbon chains of the nitrile rubber and some alkane chains have reacted with Fe. With an increase in sputtering time, the atomic concentration of C_{1S} increases, which shows that the carbon chains have the ability to permeate into the metal surface. The atomic concentration of O_{1S} is greater in the worn surface than in the original surface, which shows that the worn surface is oxidized during the test. The decrease in atomic concentration of O_{1S} with increased sputtering time means that the oxidative reaction mainly occurs on the surface of the metal.

Table 9.5. Elemental analysis of original and worn surfaces of steel T10

Steel T10		Atomic concentration (%)		
		C_{1S}	O_{1S}	$Fe_{2P2/3}$
Original surface		82.12	14.90	0.75
Worn surface	No sputtering	83.03	16.50	0.47
	2min sputtering	88.40	9.77	1.82

The XPS analysis of the original surface of steel T10 is shown in Fig.9.12. The C may have come from contamination of the test chamber. The Fe-C peak is not evident. However, the functional groups C-C, C-O, C=O, etc are present.

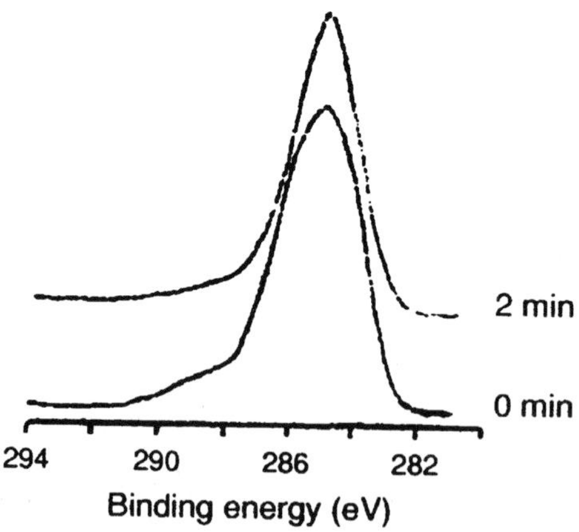

Fig.9.12. XPS spectrum of C_{1S} on the original surface of steel T10

Figure 9.13 illustrates the XPS spectra of C on the worn surfaces. The area of C_{1S} spectra for the worn surface is greater than that for the original surface. Moreover, four different chemical states, C-C (284.6eV), C-O(286.0eV), C=O(287.2eV) and

Fe-C(282.8eV) are observed (Fig.9.13). The Fe-C density is weakened after sputtering, which indicates that the reaction of carbon chains with Fe is stronger on the surface than in the subsurface. In addition, the density of C-C functional groups in the subsurface is greater than in the surface, which shows that the carbon chains are permeated into the bulk of the steel. The intensity of the higher binding energy region is stronger on the worn surface than in the subsurface, indicating that the amount of C-O and C=O in the surface is greater than in the subsurface, which means that the surface has been oxidized.

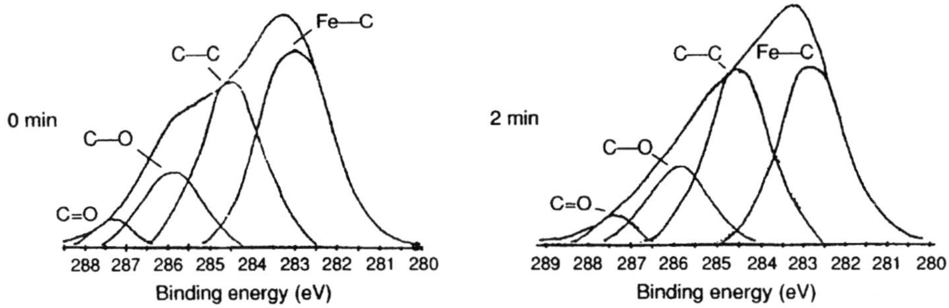

(a) No sputtering (b) 2min sputtering

Fig.9.13. XPS spectra of C_{1S} on worn surface of steel T10

9.2.3 Wear Mechanisms of Steel T10 by Nitrile Rubber [160]

9.2.3.1 Mechanochemical Reaction between Steel T10 and Alkane Molecular Chains of Mineral Oil

It appears that during the sliding process, the metal lattice was disordered and broken under the action of shear stresses. New dislocations were formed and the mechanochemical activity of metal was increased. C-C or C-H bonds in the lubricating oil were ruptured (Fig.9.8b) and produced free radicals (R), which combined with the activated metal (M) and formed a free radicals-metal (R-M) chemical adsorption films.

$$C_XH_Y + \text{Fe (metal surface)} \longrightarrow C_XH_{Y\text{-}Z}\text{-Fe (adsorption)} + \text{zH (adsorption)} \quad (1)$$

$$C_XH_{Y\text{-}Z}\text{-Fe} + O_2 \longrightarrow C_XH_{Y\text{-}Z}O \text{ (adsorption)} + \text{FeO} \quad (2)$$

The chemical reaction shown in equation (1) is one of the reasons for the increase in the carbon density and the occurrence of the Fe-C peak on the worn surface of steel T10 (Table 9.5 and Fig. 9.13). As the density of Fe-C on the surface is stronger than that in the subsurface of steel T10, this reaction mainly occurred on the surface of steel T10. The chemical reaction shown in equation (2) induces the increase in C-O and C=O on the worn surface of steel T10 (Fig.9.13).

During the experiment process, lubrication oil can be oxidized to produce substances containing C-O and C=O (Fig.9.8b), which have a catalytic action on metal. This catalytic action can trigger free radicals from hydrocarbons and cause ROOH to be resolved into free radicals, which can speed oxidation .The reaction scheme could be:

$$Fe + O_2 \longrightarrow FeO_2 \quad (3)$$

$$FeO_2 + RH \longrightarrow Fe^{2+} + R. + .OOH \quad (4)$$

$$ROOH + Fe^{2+} \longrightarrow RO. + Fe^{3+} + OH \quad (5)$$

$$ROOH + Fe^{3+} \longrightarrow RO_2 + Fe^{2+} + H^{+} \quad (6)$$

The reacting process as described in the above equations is one of the reasons for the increase in oxygen on the worn surface of steel T10 and the occurrence of iron oxides (Table 9.5).

9.2.3.2 Mechanochemical Reaction between Steel T10 and Nitrile Rubber

Under boundary lubrication conditions with mineral oil, the nitrile rubber was in contact with the oil for a long time and swelled (Fig.9.7b). Its mechanical strength is therefore significantly decreased and the other properties such as tearing resistance, wearability and aging resistance are worsened. Hence, the macromolecular chains of the nitrile rubber are ruptured easily and speedily under the action of shear stress. The decrease in CH_2 and CH_3 content of the molecular chains on the worn surface of nitrile rubber and in the used mineral oil indicates that the carbon chains of both nitrile rubber and mineral oil are broken during wear (Fig.9.8, Fig.9.9). The fracture position is at the α -CH_2 near the –CN groups, this is consistent with the FT-IR analysis for nitrile rubber (Fig.9.9) [153]. The rupture of macromolecular chains in the nitrile rubber may be illustrated as follows:

Fracture Position

$$\sim CH_2\text{-}CH{=}CH\text{-}CH_2 \;|\; CH_2 - \underset{\displaystyle CN}{\underset{|}{CH}} - CH_2 - CH = CH\sim$$

The free radical containing the –CN groups is more stable, so on the worn surface of nitrile rubber, the density of -C≡N groups increased, while the density of –CH_2- decreased, owing to the presence of free radicals from the rupture of macromolecular chains in the nitrile rubber (Fig.9.9).

Under the action of frictional force, the Fe in the surface of steel T10 is activated and can react with the macromolecular chains of nitrile rubber, provided electrons. The process could be shown as follows:

$$\sim CH{=}CH{-}CH_2{-}\underset{\displaystyle CN}{\underset{|}{CH}}{-}CH_2^{\cdot} + {}^{\cdot}Fe \xrightarrow{-e} \sim CH = CH{-}CH_2{-}\underset{\displaystyle CN}{\underset{|}{CH}}{-}H_2C - Fe \qquad (7)$$

The above reaction is confirmed by the increase in the density of Fe-C on the worn surface of steel T10 (Fig.9.13). Owing to this chemical reaction, a metal-polymer film of Fe-C was generated on the metal surface.

Since a certain amount of NBR materials were transferred to the surface of steel T10 (Fig.9.7), it is possible that the iron oxide reacted with the macromolecular chains radicals, which may be one of the reasons for the increase in carbon on the worn surface of steel T10 (Table 9.5) and for the Fe-C metal-polymer film being generated on the steel surface.

9.2.3.3 Physical Mechanisms of Wear of Steel T10 by Nitrile Rubber

As a result of the reaction processes as described in equations (2) to (6), some hard iron oxide particles are generated on the steel surface and then enter into the interface of the rubber-metal pair, which could induce the steel by microcutting, as shown in Fig.9.7(a).

Some pits were also observed on the worn surface of steel (Fig.9.7(a)), which indicates that the metallic materials were peeled off by hydrogen embrittlement under the action of alkane molecular chains from the oil. As shown in Fig.9.8, C-H and C-C bonds in molecular chains of the lubricating oil were broken and the alkane chain free radicals were generated. The free radicals reacted with the metal, and the C-M bond formed on the local fresh steel surface. Then hydrogen adsorbed at the steel surface could diffuse inside the steel, causing hydrogen embrittlement.

9.2.3.4 Summary

As stated above, the mechanisms of wear of steel by nitrile rubber under boundary lubrication conditions with mineral oil are complicated. It could be summarized as follows:

(1) A metal-polymer reaction film is produced on the steel surface by mechanochemical reaction of Fe, iron oxide, and free radicals of molecular chains from the nitrile rubber and the alkane molecular chains of the mineral oil.

(2) The metal-polymer film is damaged by microcutting acted by the iron oxide

particles.

(3) Some steel loss may result from hydrogen embrittlement.

Therefore, under boundary lubrication conditions with mineral oil, the mechanisms of wear of metal by rubber is mainly a cyclic and alternating process of generation and removal of metal-polymer films.

In addition, we also studied the wear of steel T10 by NBR, SBR, and flourorubber respectively in paraffin mineral oil with or without ZDDP. It has been found that the wear values of steel by the rubbers above are arranged in decreasing order as NBR>SBR>flourorubber in paraffin mineral oil with or without ZDDP.

9.2.4 Wear of Steel 45 by Rubber under Boundary Lubrication Condition

By using the same experimental device and method as stated above, the present author and his co-workers investigated the wear of steel 45 by NR, NBR and SBR materials respectively in water or NaOH solution. It was found that the three kinds of rubber were arranged in decreasing order of wear of steel 45 by rubber as follows: NBR>SBR>NR. The wear rate of steel is namely dependent on the media, Shore hardness and structure of macromolecular chain of the rubber.

As for the wear mechanisms of steel 45 by NR in water medium, Zhang et al [161] proposed that the wear process of steel by rubber mainly involves two stages as follows. Firstly, under the action of frictional force, some mechanochemical reactions are generated in the frictional interface, which produced a chemical reaction film on the steel surface. Secondly, this film is peeled off by microcutting of the hard particks on the sliding surface which directly leads to the wear steel. Therefore, the wear mechanism of steel 45 by rubber would be primarily attributed to the chemomechanical autocatalytic mechanism of destruction of steel being rubbed by NR material.

Chapter 10

LUBRICATION OF RUBBER SEALS

Seal is an indispensable component in the fluid machinery. A clear understanding of the lubrication and leakage mechanisms of seals is important for solving the design and operating problems of seals. Knowledge of two basic types of seal is given in this chapter based on the related basic literature. However, the main contents are introduced some research achievements reached by A. Karaszkiewicz, M. J. L. Stakenborg, H. J. Van Leenwen and E. A. M. ten Hagen as well as A. Gabelli and G. Poll [162, 164, 165, 168, 169, 172, 177].

10.1 Hydrodynamic Lubrication of O-ring Seals for Reciprocating Motion

Lubrication of seals is an important problem in the design and operation of various fluid machinery and hydraulic systems. There is a need to establish an adequate relationship defining the lubricating film thickness (leakage) for the desired seal to that end. Leakage is determined by the thickness of the lubricating film between the seal and the sealed component, such as a shaft or cylindrical bushing etc.

The problem of lubricating film thickness for rubber seals with an O-rings belongs to hydrodynamic lubrication theory for the line contact zone of low elastic modulus.

The film thickness was given as follows based on the Blok's inverse problem of hydrodynamic lubrication [163]:

$$h_c = 0.94(\eta u / p')^{0.5} \tag{10.1}$$

Where, η - dynamic viscosity of the oil;

u - velocity;

p' - value of dp/dx at inflexion of the pressure distribution curve in the lubrication film.

The above formula is an important achievement of the sealing technology in the 1960's as it can be used to solve a number of seal operating problems (leakage, friction and so forth) qualitatively. However, it has been found that considerable differences exist between theory and experiment over the whole range of ηu used in practice [164,165].

On the basis of a theoretical study of elastohydrodynamic lubrication for a highly deformed line contact zone of low elastic modulus, Hooke and O'Donoghue [166] proposed the graphical presentations of h_c and h_m (minimum lubrication film thickness) for the out- and in- strokes of the sealed element with an O-ring at a squeezing rate $\varepsilon = 0.1$. The corresponding equation for the out-stroke h_{c1} is given by [166]

$$h_{c1}/R = 2.93(\eta u/ER)^{0.6} \tag{10.2}$$

Where, E – elasticity modulus for O-ring rubber;

R – radius of ring, R=0.5d;

d – cross-section diameter of ring.

Equation (10.2) is justified theoretically under the conditions of low oil viscosities and travel velocities, i.e., $\eta u/ER < 10^{-5}$.

From the general hydrodynamic lubrication solution for the line contact zone of low elasticity modulus, the film thickness of O-rings may also be determined by [162]

$$\frac{h_m}{R} = c\left(\frac{\eta u}{ER}\right)^a\left(\frac{\omega}{ER}\right)^b \tag{10.3}$$

Where, ω - is the load contact pressure per unit perimeter of the O-ring.

Substitution of ω (for $\varepsilon = 0.1$, $\omega = 0.114ER$) into equation (10.3) as determined by Herrebrugh [167] gives

$$\frac{h_{c1}}{R} = 2.74\left(\frac{\eta u}{ER}\right)^{0.6} \tag{10.4}$$

For squeezing rate $\varepsilon = 0.1$, an experimental correlation was obtained by Karaszkiewicz [162] as follows:

$$\frac{h_{c1}}{R} = 7.77\left(\frac{\eta u}{ER}\right)^{0.71} \tag{10.5}$$

Although there are certain differences between the h_{c1}/R values taken from the theoretical equation (10.2) and the experimental equation (10.4), it has been found that equation (10.4) can qualitatively confirm the theoretical relationships obtained by including some simplifications.

Under an O-ring during an out-stroke of the sealed element, an engineering relationship defining the lubrication film thickness h_{c1} was presented by Karaszkiewicz [162]:

$$h_{c1} = 4.86(\eta u)^{0.71} E^{-0.71} R^{0.29} (2\varepsilon + 0.13)^{-0.42} \tag{10.6}$$

As seen in equation (10.6), the lubrication film thickness, h_{c1} for O-ring seals is dependent on the material and geometrical parameters (E, R, ε) as well as the operating conditions (η, u). This equation can be applied to solve certain problems in the operation and design of O-ring seals in hydraulically driven machines under various materials parameters and operating conditions.

10.2 Visco-Elastohydrodynamic (VEHD) Lubrication in Rotary Lip Seals

10.2.1 Concept of the VEHD lubrication [168,169]

In practice, a dynamic excitation of the seal lip always occurs owing to unroundnesss of the shaft or motion of shaft center. Based on a study in the influence of dynamic excitation on the occurrence of clearances in a dry seal-shaft contact, it has been shown that clearances develop due to viscous and inertial seal material behavior. These clearances will be filled with fluid in virtue of subambient cavitation pressures or various seal pumping mechanisms [170]. As the clearance geometry is no longer parallel and time dependent, a fluid film will be generated thanks to entrainment and squeeze effects under certain conditions. Reasoning in a similar way, shaft out-of-roundness and more general multi-lobed or shaft waviness will also produce similar effects and fluid film formation [168,171]. Therefore, this phenomenon is called visco-elastohydrodynamic (VEHD) lubrication. It is a type of full film lubrication resulted from the viscous and inertial effects of seal material behavior.

10.2.2 Interaction between the Viscoelastic Seal Deformation and the Fluid Film Formation [168,169]

For a whirling shaft, it is assumed that the shaft surface is a perfect smooth cylinder and the shaft center C_{sh} rotates around the seal center C_{se} at whirl angular velocity ω_w as shown in Fig. 10.1 The seal has no microasperities, and the deformation is described by the linerized transfer function as follows [168]:

$$H(f, y_i) = \frac{P(f, y_i)}{X(f, y_i)} = \frac{P(f, y_i)}{X(f)} \tag{10.7}$$

Where, $P(f, y_i)$ - a complex variable representing the Fourier transform of

$p_t(t, y_i)$ at node i;

$X(f)$ - Fourier transform of $x_t(t)$;

$p_t(t, y_i)$ - total contact stress at node i.

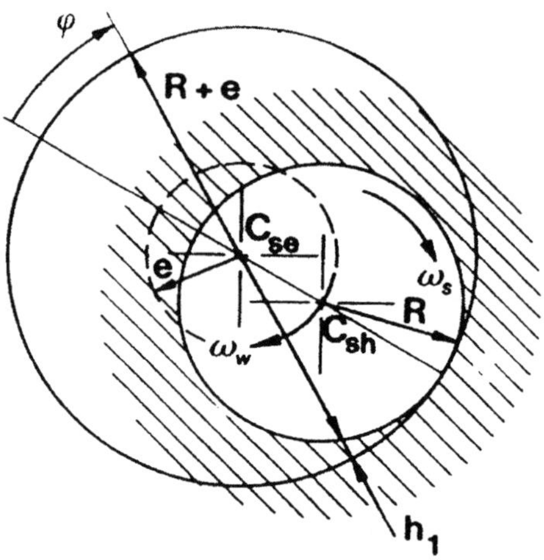

Fig.10.1. Harmonic seal excitation due to radial shaft run out without fluid film effects [169]

The seal is forced to perform oscillations around a stretched state at radius $R+e$ (Fig.10.1). Moreover, a material point at the seal circumference will experience a harmonic motion if it remains in contact with the shaft.

It is assumed that the fully flooded conditions are in the clearances. The relationship between the fluid film pressure and the film geometry is given by Reynold's equation for Newtonian fluids. If the shaft and seal are coaxial, the equation can be given as follows in cylindrical coordinates [169]:

$$\frac{1}{R^2}\frac{\partial}{\partial\theta}\left(\frac{h^3}{12\eta}\frac{\partial p}{\partial\theta}\right)+\frac{1}{b^2}\frac{\partial}{\partial y}\left(\frac{h^3}{12\eta}\frac{\partial p}{\partial y}\right)=\frac{1}{2}\left(\omega_s+\omega_{se}\right)\frac{\partial h}{\partial\theta}+\frac{\partial h}{\partial t} \tag{10.8}$$

Where,

R – shaft radius;

b – seal-shaft contact width;

h – film thickness;

p – fluid film pressure;

t – time;

y – axial coordinate in contact area;

η - dynamic viscosity;

θ - circumferential coordinate;

ω_s - shaft angular velocity;

ω_{se} - seal angular velocity.

In equation (10.8), the stretching effects and the influence of fluid shear stresses on the seal deformation are without consideration.

For a radial lip seal b/R is of the order of 10^{-3}. Based on a short bearing model and the introduction of a rotating reference system, the pressure distribution p_0 can be expressed by [169]:

$$\frac{\partial}{\partial y}\left(\frac{h^3}{12\eta}\frac{\partial p_0}{\partial y}\right)=\frac{1}{2}\left(\omega_s+\omega_{se}-2\omega_{ref}\right)\frac{\partial h}{\partial \theta}+\frac{\partial h}{\partial t} \tag{10.9}$$

Where ω_{ref} is the angular velocity of the reference system.

For the angular velocity of the reference system, two convenient choices are (a) $\omega_{ref}=\omega_w$ (whirl angular velocity), a stationary bearing problem with entrainment effects only, or (b) $\omega_{ref}=(\omega_s+\omega_{se})/2$, a dynamically loaded bearing problem with squeeze effects only. However, both cases above produce the same pressure distribution and film thickness profile. In this case, $\omega_{se}=0$. Presuming $h\neq h(y)$, so shaft and seal are parallel in axial direction. Then for case (a), equation (10.9) is given by

$$\frac{\partial}{\partial y}\left(\frac{h^3}{12\eta}\frac{\partial p_0}{\partial y}\right)=-\left(\omega_w-\frac{\omega_s}{2}\right)\frac{\partial h}{\partial \phi} \tag{10.10}$$

Where $\phi=\theta+\omega_w\cdot t$.

Presuming $p_0(\phi, y=0)=p_0(\phi, y=b)=0$, ambient pressure is represented by $p=0$, then equation (10.10) can be integrated as follows:

$$p_0(\phi, y)-A\cdot\frac{\frac{dh}{d\phi}}{h^3}=0 \tag{10.11}$$

Where

$$A=-6\eta b^2\left(\omega_w-\frac{\omega_s}{2}\right)\frac{y}{b}\left(\frac{y}{b}-1\right) \tag{10.12}$$

The pressure distribution p_0 rotates along the seal surface at frequency f_w. According to equation (10.11), a material point at $y=y_3$ in the contact area experiences the time dependent pressure, where $\phi = 2\pi f_w t$. The pressure will separate the preloaded seal from the draft with radius of *R*+*e*. It is given by

$$p_d(y_3,t) = p_0(y_3,t) - p_s(y_3) \tag{10.13}$$

Where, $p_s(y_3)$ represents the static contact stress for a seal stretched at a radius of *R*+*e*.

Based on the analysis of dynamic seal behavior [168], the seal displacement response $\delta(t,y)$ due to the fluid film pressure signal $p(t,y)$ can be given by:

$$\delta(t,y) = \int_{-\infty}^{\infty} g(t-\tau,y)\cdot p(\tau,y)d\tau \tag{10.14}$$

Where, τ is time integration variable.

Assuming that both the viscoelastic displacement response $\delta(y_3,t)$ and $p_d(y_3,t)$ are periodic with period $T = 1/f_w$, equation (10.14) can be expressed as [169]:

$$\delta(y,t) = \delta(y,t+nT) = \int_0^T g(y,t-\tau)\cdot p_d(y,\tau)d\tau \tag{10.15}$$

Where, *n* is the discrete number of periods. Compliance function $g(y,t)$ includes the inertia of garter spring and seal material.

Now, the total film thickness can be written as [169]:

$$\begin{aligned} h(y_3,t) &= h(y_3,t+nT) = h_1(t) + \delta(y_3,t) \\ &= e.[1+\cos(2\pi f_w t)] + \delta(y_3,t) \end{aligned} \tag{10.16}$$

Where, h_1 - clearance for seal stretched at radius *R*+*e* (Fig.10.1);

e - eccentricity between C_{se} and C_{sh};

f_w - whirl frequency.

To sum up, the theory of the VEHD lubrication may offer an explanation for the circumferentially nonparallel film geometry in a radial lip seal.

10.3 Micro-Elastohydrodynamic (MEHD) Lubrication in Rotary Lip Seals

10.3.1 MEHD Film Modeling [172]

The term of micro-elastohydrodynamic (MEHD) lubrication was introduced by Christensen in 1967 [173]. This lubrication phenomenon deals with the local pressure and film thickness fluctuations developed at the contacting spots of colliding asperities. Later on, Tsao and Tong [174] introduced further this idea into the situation of parallel sliding surfaces using an elementary model to allow the asperities to gain contact and to deform. However, the roughness was represented by an oversimplified geometry consisting of a short cylinder with a spherical top.

It has been proved that the film formation of rotary seal lubrication is arised from a combined MEHD and hydrodynamic (HD) action of the surface asperities interacting at the seal interface [175-177]. Based on the above conclusion, Gabelli and Poll [172] introduced the effect of rubber viscoelasticity and squeeze film into the film model. They considered that the load-carrying capacity of the parallel rough sliding surfaces in rotary contacts of seals is combined of three components as follows:

(1) Hydrodynamic Component

It is resulted from the integration of the HD pressure field developed by the wavy morphology formed by the combined effect of the roughness present on both surfaces of the sealing contact.

(2) Microcontact load component

It is derived by the actual contact of the asperities' summits of the sealing surfaces. The low elastic modulus of the rubber ensures that collicing asperities will be separated by the iso-viscous EHD lubricant film in all but the most difficult situations.

(3) Squeeze-Film component

It accounts for the load support provided by the oscillatory motion of the sealing counterface and the consequent dynamic response of the rubber lip.

10.3.2 Hydrodynamic Analysis [172]

Considering the laminar flow of an incompressible Newtonian lubricant in the lip sealing interface, the shape of the sealing gap is related to the morphology of the two surfaces in contact for moderate load cases. The seal lip junction might be considered as a contact of an equivalent rough surface (seal) with an ideally smooth plane (shaft) since the rubber surface is much rougher than the steel surface. Thus, the Reynolds' equation governing the 2-D hydrodynamic pressure field $p(x, y)$ in the sealing junction can be given by [172]

$$\frac{\partial}{\partial x}\left(h^3 \frac{\partial p}{\partial x}\right)+\frac{\partial}{\partial y}\left(h^3 \frac{\partial p}{\partial y}\right)=6\eta_0 u_x \frac{\partial h}{\partial x} \tag{10.17}$$

Where,

h – lubricant film thickness;

u_x – absolute sliding speed;

η_0 – dynamic viscosity at atmosphere pressure.

By normalizing the equation above, we have

$$\frac{\partial}{\partial x}\left(H^3\frac{\partial p}{\partial x}\right)+\frac{\partial}{\partial Y}\left(h^3\frac{\partial P}{\partial Y}\right)=6\frac{\partial H}{\partial X} \tag{10.18}$$

Where *H, P, X* and *Y* are dimensionless groups, namely

$$H=\frac{h(x,y)}{\sigma},\quad P=\frac{\sigma^2 p}{\eta_0 u_x b},\quad X=\frac{x}{b},\quad Y=\frac{y}{b} \tag{10.19}$$

Where b is the seal lip contact width, σ is standard deviation of height distribution.

Based on the results of numerical analysis [177], the load-carrying capacity (hydrodynamic pressure) developed in the lip sealing interface is given by

$$P_h=\frac{1}{A_0}\iint_{A_h}p(x,y)dxdy \tag{10.20}$$

And

$$A_h=A_0-A_c-A_r \tag{10.21}$$

Where, A_h - area of the hydrodynamic field;

A_0 - apparent unit area of contact;

A_c - area of cavitation;

A_r - real area of microcontact.

If the pressure distribution is known, the mean shear stress resulted from the hydrodynamic field can be derived [172].

10.3.3 Microcontact Analysis [172]

In the analysis of two nominally flat rough surfaces representing the seal-shaft junction [172], the load, pressure and geometry of the contacting spots generated in the sealing interface can be calculated by applying the Greenwood-Tip theory [178]. The apparent pressure acting on the microcontact is:

$$p_r(\underline{h})=24\times10^{-4}m_2^{-2}E'\left(\frac{m_0m_4(\alpha-0.8968)}{\alpha}\right)^{\frac{5}{4}}\Omega_{2.5}(\underline{h}) \tag{10.22}$$

Where, $\underline{h}$ - dimensionless separation;

m_0, m_2 and m_4 - roughness spectral moment of order 1, 2, 4 respectively;

E' - composite plan strain modulus of elasticity;

α - bandwidth parameter, $\alpha=m_0m_4m_2^{-2}$;

$\Omega_{2.5}(\underline{h})$ - microcontact expectancy function of order 2.5.

And the real force per contact spot is:

$$F_r(\underline{h}) = 1.85E'\left(\frac{m_0 m_4(\alpha - 0.8968)}{\alpha}\right)^{\frac{1}{4}} \Omega_{2.5}(\underline{h})\Omega_2^{-1}(\underline{h}) \qquad (10.23)$$

$\Omega_{2.5}(\underline{h})$ and $\Omega_2(\underline{h})$ are the Probability Density Function (PDF) of the Summits' height or the probability associated with a randomly selected summit having an asperity to asperity contact, namely:

$$\Omega_q(\underline{h}) = \int_h^{\infty} (s - h)^q \phi(s) ds \qquad (10.24)$$

Where $\phi(s)$ is the normalized or standard frequency density function to describe the asperities' height distribution and s is a dummy variable of integration.

m_0, m_2 and m_4 in equations (10.20) and (10.21) are the spectral moments of the Power Spectral Density Function (PSDF) $G(\omega)$ of the two surfaces, namely:

$$m_n = (2\pi)^n \int_{\omega_1}^{\omega_2} \omega^n G(\omega) d\omega \qquad (10.25)$$

Where ω_1 and ω_2 are the lower and upper spatial frequencies of the roughness profile.

Figure 10.2 makes a comparison of the parametric computation of the load support of sealing contacts between the computations performed using equation (10.20) and the microcontact calculation performed in accordance with equation (10.22). As shown, even at low rotary speeds, the contribution of the asperities' collision is very small. However, in the case of oil starvation in the sealing junction, the microcontact load support can be critical for the survival of the sealing element.

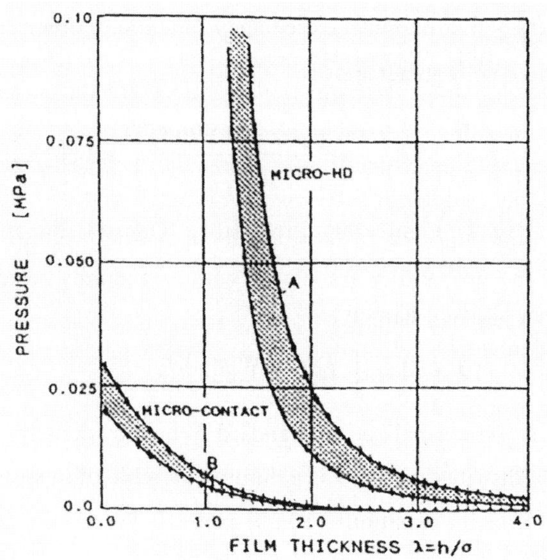

Fig.10.2 Micro-hydrodynamic lubrication model, relationship between specific load-carrying capacity, i.e., pressure and lubricant film thickness [172]:
(A)Global load –carrying capacity; (B) Load carrying of asperities' contact.

10.3.4 Squeeze-Film Analysis

A general oscillatory motion of the seal contact and a dynamic variation of the pressure in the sealing junction can be induced by the small geometrical imperfections of the rotating shaft forming the sealing counterface. The amplitude and phase shift of the pressure variation will depend on the parameters characterizing the visco-elastic response of the rubber lip. A quick quantification of the increased load-carrying capacity provided by this dynamic effect can be obtained by first computing the changes of the contacting pressure and then the variation of the lubricant film thickness, and the carrying capacity provided by the induced squeeze action of the lubricant film [172].

On the basis of a Grubin type of analysis [179], the squeeze-film pressure component induced by the periodic contact force of the seal lip can be obtained:

$$p_s = -\frac{\eta_0 b^2}{h^3}\frac{dh}{dt} \tag{10.26}$$

From the equilibrium condition of the various forces acting at the seal lip contact, the acting average pressure of sealing contact can be given by [172]:

$$P(\theta,t) = p_h + p_r + p_s = p_h + p_r - \frac{\eta_0 b^2}{h^3}\frac{dh}{dt} \tag{10.27}$$

Equation above can be solved for a time interval corresponding to the minimum and maximum points of the film thickness.

Comparing the film thickness predicted using the model presented in this section with the experimentally measured film thickness using magnetic fluid [180], it was found that the correlation is good for the full range of speeds tested. However, the validity of the model is constrained to those speeds beyond which the hydrodynamics of the system becomes strongly coupled with the macro-elastodynamics of the seal lip deformation.

References

[1] Department of Education and Science, Lubrication (Tribology) Education and Research. A Report on the Present Position and Industry's Needs, Department of Education and Science, HMSO, London, 1966.
[2] H. Czichos, Tribology – A System Approach to the Science and Technology of Friction. Lubrication and Wear, Elsevier, Amsterdam, 1978.
[3] Siwei Zhang, Principal and methodology of systems engineering for tribology, Lubrication and Seals 6(1980)7 (in Chinese).
[4] Siwei Zhang, Chinese J. Mechanical Engineering (English Edition) 6(1993)40.
[5] W.A. Glaeser, Materials for Tribology, Elsevier, Amsterdam, 1992.
[6] C. Kajdas, S.S.K. Harvey and E. Wilusz, Encyclopedia of Tribology, Tribology Series 15, Amsterdam, Elsevier, 1990.
[7] I.V. Kragelsky and V.V. Alisin, Tribology-Lubrication, Friction and Wear. Tribology in Practice Series, Eds. M.J. Neale, T.A. Polak and C.M. Taylor, Professional Engineering Publishing Ltd., London, 2001.
[8] F.L. Roth, R.L. Driscoll and W.L. Holt, J. Res. Natl. Bur. Stand, 28(1942)439.
[9] R.P. Cooper and B.Ellis, J. Appl. Polym. Sci., 27(1982)4735.
[10] B. Ellis, Wear, 93(1984)111.
[11] S.W. Zhang, Examination of the effect of run-in on rubber friction, Proc. 5th Int. Congr. on Tribology, Eds. K. Holmberg and I. Nieminen, Espoo, Finland, 1989, v.3, p341.
[12] D. Tabor, Wear, 1(1957/1958)5.
[13] D.F. Moore, The Friction and Lubrication of Elastomers, Pergamon Press, Oxford, 1972.
[14] M.W. Pascoe and D. Tabor, Proc. Roy. Soc., A, 235(1956)210.
[15] A.R. Savkoor, Wear, 8(1965)222.
[16] M.R. Hatfield and G.B. Rathmann, J. Phys. Chem. 60(1956)957.
[17] A. Schallamach, Wear, 1(1957/1958)384.
[18] A. Schallamach, Proc. Phys. Soc. B 67(1954)883.

[19] D.F. Moore, Principles and Application of Tribology, Pergamon Press, Oxford, 1975.
[20] K.R. Eldredge and D. Tabor, Proc. Roy. Soc. A 229(1955)181.
[21] Xu Hao, Machine Design Handbook, Vol.1, Machine Building Press, Beijing, 1991 (in Chinese).
[22] C.S. Wilkinson, Rubber. Chem. Technol. 27(1954)255.
[23] K.A. Grosch, Proc. Roy. Soc. Lond., A274(1963)21.
[24] E. Southern and R.W. Walker, Nature, 237(1972)142.
[25] A. Schallamach, Wear, 71(1971)301.
[26] A.D. Roberts and S.A. Jackson, Nature, 257(1975)118.
[27] A.D. Roberts and J.R. Richardson, Wear, 67(1981)55.
[28] A.D. Roberts, Tribol. Int., 14(1981)14.
[29] A.D. Roberts, J. Nat. Rubber. Res., 2(1987)255.
[30] R.A. Moyer, Proc. 27th Annual Meeting, Washington D.C., 1947, p.346.
[31] J.E. Footit, SAE 770279, 1977.
[32] R.J. William, SAE 800839, 1980.
[33] J.E. Humter, SAE 930896, 1993.
[34] R. Eddie, SAE 940724, 1994.
[35] N. Francis, M. Michael and N. Connie, SAE 960652, 1996.
[36] D.P. Martin and G.F. Schaefer, SAE 960657, 1996.
[37] G.F. Hayhoe and C.G. Shapley, SAE 890028, 1989.
[38] K. Shimizhu and C Ikeya, SAE 890004, 1989.
[39] K. Shimizhu, SAE 910167, 1991.
[40] K. Shimizhu and M. Nihei, SAE 920018, 1992.
[41] X.D. Peng, Y.B. Xie and K.H. Guo, SAE 990478, 1999.
[42] X.D. Peng, Y.B. Xie and K.H. Guo, SAE 001640, 2000.
[43] K. Naab and R. Hoppstock, Sensor Systems and Signal Processing for Advanced Driver Assistance. In: Smart Vehicles, Eds. J.P. Pauwelussen and H.D. Pacejke, Swets & Zeitlinger, 1996, p.69.
[44] M.B. Peterson and W.O. Winter, Wear Control Handbook, New York, ASME, 1980.
[45] J.T. Burwell, Wear, 1(1957/1958)119.
[46] D.F. Moore, Some observations on the interrelationship of the friction and wear in elastomers, The Wear of Non-metallic Materials, Eds. D. Dowson, M. Godet and C.M. Taylor, London, Mechanical Engineering Publication Ltd., 1978, p.141.
[47] D.F. Moore, Wear, 61(1980)273.
[48] A.Schallamach, Trans. Inst. Rubber Ind., 28(1952)256.
[49] A.Schallamach, Int. Polym. Sci. and Technol., 6(1979)T/44.
[50] S.W. Zhang, Tribol. Int. 22(1989)143.
[51] A.Schallamach, J. Polym. Sci., 9(1952)385.

[52] L. Ahman and A. Oberg, Wear of Materials 1983, Ed. K.C. Ludema, ASME, 1983, p.112.
[53] D.H. Champ, E. Southern and A.C. Tomas, in Advance in Polymer Friction and Wear, Ed. L.H. Lee, New York, Plenum Press, 1974. p.133.
[54] E. Southern and A.C. Thomas, Rubber Chem. Technol., 52(1979)1008.
[55] A.N. Gent and C.T.R.Pulford, Appl. Polym. Sci., 28(1983)43.
[56] S.W. Zhang, Rubber Chem. Technol. 57(1984)755.
[57] S.W. Zhang, Rubber Chem. Technol. 57(1984)769.
[58] S.W. Zhang, in Polymer Wear and its Control, Ed. L.H. Lee, ACS, Washington D.C., 1985, p.189.
[59] V. I. Dyrda, V.I. Vettegren and V.P. Nadutgi, Int. Polym. Sci. Technol., 3(1976)T/66.
[60] Y. Uchiyama and Y. Ishino, Wear, 158(1992)141.
[61] Y. Fukahori and H. Yamazaki, Wear, 171(1994)195.
[62] Y. Fukahori and H. Yamazaki, Wear, 178(1994)109.
[63] Y. Fukahori and H. Yamazaki, Wear, 188(1995)19.
[64] E. Southern and A.G. Thomas, in The Wear of Non-metallic Materials, Eds. D. Dowson, M. Godet and C.M. Taylor, London, Mechanical Engineering Publication Ltd., 1978, p.157.
[65] A.G. Thomas, J. Polym. Sci. Symp., 48(1974)145.
[66] D.H. Champ, E. Southern and A.G. Thomas, Am. Chem. Soc., Coatings Plast. Div. Prepr., 34(1974)237.
[67] A.K. Bhowmick, S. Basu and S.K. De, J. Mater. Sci., 16(1981)1654.
[68] A.K. Bhowmick, S. Basu and S.K. De, Rubber Chem. Technol., 55(1982)23.
[69] A.K. Bhowmick, and S.K. De, Rubber Chem. Technol., 52(1979)985.
[70] A.K. Bhowmick, and S.K. De, Rubber Chem. Technol., 53(1980)960.
[71] A.K. Bhowmick, G.B. Nanda, S. Basu and S.K. De, Rubber Chem. Technol., 53(1980)327.
[72] A.K. Bhowmick, Rubber Chem. Technol., 55(1982)1055.
[73] S.K. Chakraborth, A.K. Bhowmick and S.K. De, Rubber Chem. Technol., 55(1982)41.
[74] S.K. Chakraborth and S.K. De, J. Appl. Polym. Sci., 27(1982)4561.
[75] B. Kuriakose and S.K. De, J. Mater. Sci., 20(1985)1864.
[76] S. Thomas, B. Kuriakose, B.R. Gupta and S.K. De, J. Mater. Sci., 21(1986)711.
[77] S.Thomas, Wear, 116(1987)201.
[78] J.A. Schweitz and L. Ahman, Wear of Materials 1983, Ed. K.C. Ludema, ASME, 1983, 610.
[79] L. Ahman, A. Bachlin and J.A. Schweitz, Wear of Materials, 1981, Eds. S.K. Rhee, A.W. Ruff and K.C. Ludema, ASME, 1981,161.
[80] V.I. Dyrda, V.I. Vettegren and V.P. Nadutyi, Int. Polym. Sci. and Technol., 3(1976)66.

[81] P. Thavamani, D. Khastgie and A.K. Bhowmick, J. Mater. Sci., 28(1993)6318.
[82] I.A. Abu-Isa, Rubber Chem. Techol., 8(1985)867.
[83] A. Schallamach, in Chemistry and Physics of Rubber-like Substances, Ed. L. Bateman, London, MacLaren, 1963, p.355.
[84] A. Schallamach, Rubber Chem. Technol., 41(1968)209.
[85] K.A. Grosch and A. Schallamach, Trans. Inst. Rubber Ind. 41(1965)T80.
[86] K.A. Grosch and A. Schallamach, Kaut. u. Gummi, Kunstst., 22(1969)288 (in German).
[87] D.I. James, Ed., Abrasion of Rubber, McLaren & Sons, Ltd., London, 1967.
[88] A.N. Gent and C.T.R. Pulford, in Developments in Polymer Fracture-1, Ed. E.H. Andrews, London, Applied Science Publishers Ltd., 1979, p.155.
[89] A. Schallamach, J. Appl. Polym. Sci., 12(1968)281.
[90] G.I. Brodskii, et al., Soviet Rubber Technol., 8(1960)22.
[91] Y. Uchiyama, Wear, 110(1986)369.
[92] B.B. Boonstra, F.A. Heckman and Ya A. Kaba, Rubber Age, 104(1972)33.
[93] A.D. Sarkar, Friction and Wear, London, Academic Press Inc., 1980, p.267.
[94] A.N. Gent and C.T.R. Pulford, Wear, 49(1978)135.
[95] A.N. Gent and C.T.R. Pulford, J. Mater. Sci., 14(1979)1301.
[96] A.N. Gent and W.R. Rodgers, J. Polym. Sci. Polym. Chem. Ed., 23(1985)829.
[97] A.N. Gent, Proc. Int. Conf. on Rubber and Rubber-like Materials, Jamshedpur, India, 6-8, Nov., 1986,p.78.
[98] A.H. Muhr and A.G. Thomas, Proc. Int. Conf. on Rubber and Rubber-like Materials, Jamshedpur. India, 6-8, Nov., 1986, p.68.
[99] C.T.R. Pulford, J. Appl. Polym. Sci., 28(1983)709.
[100] A.H. Muhr and A.D. Reberts, Wear, 158(1992)213.
[101] S.W. Zhang, Chang Xibin, Proc. Japan Int. Tribology Conf., Nagoya, Oct. 29-Nov.1, 1990, JST, Osaki Printing Co. Ltd., 1990, 249.
[102] S.W. Zhang, Wear, 158(1992)1.
[103] B.Burr and K.M. Marshek, Wear, 81(1982)347.
[104] P.B. Lindley and A.G. Thomas, Proc. 4th Rubber Technol. Conf. London, IRI, London, 1962,p.428.
[105] A.N. Gent, Rubber Chem. Technol., 62(1988)750.
[106] O.J. Smith and C.K. Liu, ASME Trans., 75(1953)157.
[107] G. Fleischer, Schmierungstechnik, 4(1973)9 (in German).
[108] S.W. Zhang and Zhaochun Yang, Tribol. Int., 30(1997)839.
[109] P.R. Stupak and J.A. Donovan, J. Mater. Sci., 23(1988)2230.
[110] P.R. Stupak, J.H. Kang and J.A. Donovan, Wear, 141(1990)73.
[111] P.R. Stupak, J.H. Kang and J.A. Donovan, Mater. Characterization, 27(1991)231.
[112] Xie Heping and D.J. Sanderson, Fractal Effect of Dynamic Crack Extension, ACTA Mechanica Sinica, 27(1995)18 (in Chinese).

[113] B. Dubuc, J.F. Quiniou, C.Carmes, C. Tricot and S.W. Zucker, Phys. Rev. A 39(1989)1500.
[114] I.V. Kragelsky, M.N. Dobychin and V.S. Kombalov, Friction and Wear Calculation Methods, N. Standon, (trans.) Pergamon Press, Oxford, 1982.
[115] S.F. Scieszka, Tribol. Trans., 35(1992)59.
[116] S.W. Zhang, Wang Deguo, Yin Weihua, J. Mater. Sci., 30(1995)4561.
[117] S.W. Zhang, J. Eng. Tribol. Part. J., 212(1998)227.
[118] S.W. Zhang, Renyang He, Deguo Wang. Qiyun Fan, J. Mater. Sci., 36(2001)5037.
[119] I.M. Hutchings, Tribology: Friction and Wear of Engineering Materials, Edward Arnold, London, 1992.
[120] I. K. Libijef, Erosion in boiler units and its prevention, Power Station, 11(1958)22 (in Russian).
[121] I. Finnie, Wear, 3(1960)87.
[122] V.H. Blachikov, Erosion in pipe surfaces of boiler units, Bulletin of Institutes of Higher Learning, 1(1958)94 (in Russian).
[123] J.G.A. Bitter, Wear, 6(1963)5.
[124] J.H. Neilson, A. Gilchrist, Wear, 11(1968)111.
[125] Wang Deguo, Investigation of mechanisms of abrasive erosion of polymers [Master Thesis], Beijing, University of Petroleum, 1991 (in Chinese)
[126] E.E. Stalel, S.B. Latnel, Ways to increase the erosion resistance of polymeric components, Bulletin Manufacture, 5(1971)34 (in Russian).
[127] M.C. Roco, G.R. Addie, Power Technol. 50(1987)35.
[128] M.C. Roco, P. Nair, G.R. Addie, J. Dennis, J. Pipelines, 4(1984)213.
[129] M.C. Roco, C.A. Shook, J. Fluids Engng., 107(1985)224.
[130] M.C. Roco, P. Nair, G.R. Addie, J. Fluids Engng., 108(1986)455.
[131] M.C. Roco, P. Nair, Testing Apparatus for Slurry Erosion Wear, Report IMMR DOE, Univ. Kentucky, 1983.
[132] J.C. Arnold, I.M. Hutchings, Wear, 138(1990)33.
[133] J.C. Arnold, I.M. Hutchings, Phys. D: Appl. Phys., 25(1992)A222.
[134] S.W. Zhang, Yang Zhaochun, Proc. Japan Int. Tribology Conf., Yokohoma, Han Lim Won Publishing Co., Korea, 1996, 217.
[135] I.V. Kragelsky, Wear, 8(1965)303.
[136] S. Bahadur, Wear, 60(1980)237.
[137] M.K. Kar, Wear, 63(1980)105.
[138] I.V. Kragelskii, Soviet Rubber Technol., 18(1959)22.
[139] M.M. Reznikovskii, K.H. Lazareva, Fatigue characteristics of rubbers under the action of cyclic loading, Rubber, 3(1963)17 (in Russian).
[140] M.M. Reznikovskii, Soviet Rubber Technol., 19(1960)30.
[141] I.V. Kragelskii, E.F. Nepomnyaschchii, Wear, 8(1965)303.
[142] G.E. Blodski, V.F. Efstlatova, N.L. Sahanofski, L.D. Slujkov, Wear of Rubber,

Moscow Chemistry Press, 1975 (in Russian).
[143] R.S. Rivlin, A.G. Thomas, J. Polym. Sci., 10(1953)291.
[144] H.W. Greensmith, Polym. Sci., 21(1956)175.
[145]W. Ostwald, Handbuch der Allgemeinen Chemie, I. Akad. Verlagsanstalt, Leipzig, 1919, p.70 (in German).
[146] P.M. Travis, Mechanochemistry and the Colloid Mill Including the Practical Applications of Fine Dispersion, New York, The Chemical Catalog Co. Inc, 1928.
[147] G. Heinicke, Tribochemistry, Akademic-Verlag, Berlin, 1984.
[148] R. Porter, A. Casale, Encycl. Polym. Sci. Eng., J.L. Kroshwitz (Ed.), Willey, 1987.
[149] E.M. Gutman, Mechanochemistry of Materials, Cambridge, Combridge International Science Publishing, 1998.
[150] N.K. Baramboim, Mechanochemistry of Polymers, London, Maclaren Sons Ltd., 1964.
[151] G. W. Rowe, Friction Wear and Lubrication Terms and Definitions, Organization for Economic Cooperation and Development, Birmingham, Gt. Britain, 1966.
[152] J.A. Brydson, Rubber Chemistry, London, Applied Science Publisher Ltd., 1978.
[153] A. Casale and R.S. Porter, Polymer Stress Reactions, Vol.1, Academic Press Inc., New York, 1978.
[154] B. Vollmert, Polymer Chemistry, Heidelberg, New York, 1973.
[155] A.V. Tobolsky, H.F. Mark, Eds., Polymer Science and Materials, John Wiley, New York, 1971.
[156] J. Halling, Principles of Tribology, The Macmillan Press, New York, 1975.
[157] R.B. King and J.K. Lancaster, Wear, 61(1980)341.
[158] K.AB-Malek and A. Stevenson, J. Mater. Sci., 19(1984)585.
[159] J.M. Charrier, S.G. Maki, J.P. Chalsfoux and A.N. Gent, J. Elastomers and Plastics, 18(1986)200.
[160] S.W. Zhang, Liu Qiongjun and He Renyang, Lubrication Science, 13(2001)167.
[161] S.W. Zhang, Liu Haichun and He Renyang, Wear, 256(2004)226.
[162] A. Karaszkiewicz, Tribol. Int., 21(1988)361.
[163] H. Blok, Inverse problems in hydrodynamic lubrication and design directives for lubricated flexible surfaces, Symp. on Lubrication and Wear, University of Houston, TX, USA, 1963.
[164] A. Karaszkiewicz, Hydrodynamics of rubber seals for reciprocating motion, 10th Int. Conf. on Fluid Sealing, Cranfield, U.K., 1984, paper F2.
[165] A. Karaszkiewicz, I & EC Product R/D (ACS), 24(1985)283.
[166] C.J. Hooke and J.P. O'Donoghue, J. Mech. Eng. Sci., 14(1972)34.
[167] K. Herrebrugh, J. Lubr. Technol., (1968)262.
[168] M.J.L. Stakenborg, H.J. van Leeuwen and E.A.M. ten Hagen, Trans. of the ASME, 112(1990) 578.
[169] H.J. van Leeuwen and M.J.L. Stakenborg, Trans. of the ASME, 112(1990)584.

[170] H.K. Mueller, Concepts of sealing mechanism of rubber lip type rotary shaft seals, Proc. 11th Int. Conf. on Fluid Sealing, Cannes, France, Apr. 8-10, 1987, B.S-Nau (ed.), Elsevier, London, Paper K1, pp.698-709.
[171] J.D. Symons, ASME J. Lubr. Technol., 90(1968)365.
[172] A. Gabelli and G. Poll, Trans. of the ASME, 114(1992)280.
[173] H.Christensen, Micro-elastohydrodynamics, Chapter 4 of Tech. Rep. MIT 67TR23, May 1967.
[174] Y.H. Tsao and K.N. Tong, Trans. of the ASLE, 18(1975)96.
[175] T. Ishii, S. Miyagawa and I. Taguma, Trans. of the ASLE, 29(1985)339.
[176] M. Ogata, T. Fujii and Y. Shimotsuma, Proc. of the 14th Leeds-Lyon Symp. on Tribology, 1987, p.553.
[177] A. Gabelli, Proc. of the 15th Leeds-Lyon Symp. on Tribology, 1988, p. 57.
[178] A.J. Greenwood and J.H. Tripp, Proc. Inst. of Mech. Engrs., 185(1970)625.
[179] G.R. Higginson, Wear, 46(1978)387.
[180] G. Poll and A. Gabelli, Trans. of the ASME, 114(1992)290.

Subject Index

Abradibility, 36, 61
  see also energetic wear rate
Abrasion, 37, 39
  factor, 36, 43
  pattern, 39
    formation, 92
    origin of, 88
    overlap levels of, 123
    spacings of, 48
  see also abrasive wear
Abrasive (sliding) erosion, 135
  test machine, 136
  wear speed, 146
    influence factors, 148
    wear equation, 150
  worn surface characteristics and physical processes of, 137
    fluoroelastomer (FE), 144
    natural rubber (NR), 141
    nitrile rubber (NBR), 143
    polyurethane (PU), 137
    styrene-butadiene (SBR), 139

Coefficient
  of abrasion resistance, 36, 43
    see also wearability
  of friction, 9, 17, 21, 23, 27
    of adhesion, 63
    of hysteresis, 63
  of wear, 37
    see also wear constant

Day abrasion, 39
  line contact, 39
    linear wear-rate equation, 51
      wear mechanism, 46
  multiple-point contact, 55
    rates of wear, 58
    wear mechanism, 55
  point contact, 39

Elastomers, 1
  basic features, 5
  concept, 1
  definition, 2
Energetic wear rate, 36
  see also abradibility
Energy theory of metal wear, 100
  basic energy density, 101
  hypothetic frictional energy density, 101

Fatigue wear, 37, 177
fatigue resistance, 178
rate of wear, 178
wear mechanism, 177
Fractal analysis, 110
fractal dimension, 110
debris, 113
wear surface, 110,115, 119
of wear debris, 113
computer image analysis, 126
index of abradibility, 128
of wear (worn) surface ,110
calculation of crack angle
of abrasion pattern, 113
computer simulation, 123
variation method, 115
Friction, 7
definition, 7
essential characters, 7
forces on ice, 28
rubber-like materials
on ice, 28
tire traction on ice, 30
of rubber, 24
by a line contact, 24
by a point contact, 14
Frictional wear, 38, 177, 180
rate of wear, 183
wear mechanism, 180
see also wear caused byroll formation

Gravimetric wear rate 36
of erosion, 135

Hydrodynamic lubrication, 247

Impact erosion, 135
Intrinsic abrasion, 39, 61

Linear wear rate, 36
of rubber abrasion, 51, 59
line contact, 51
multiple-point contact, 58
Lubrication of rubber seals, 247
O-ring seals, 247
rotary lip seals, 249, 252

Mechanochemistry, 185
Micro-elastohydrodynamic (MEHD)
Lubrication, 252
film modeling, 252
hydrodynamic analysis, 253
microcontact analysis, 254
squeeze-film analysis, 256

Oily abrasion, 39, 66
basic features, 66
wear behavior, 67

Pattern abrasion, 39

Relative wear resistance, 37
Run-in friction, 8
characteristics, 8
effect of initial contact state, 11
hypothesis on the
physical processes, 12

Surfacial mechanochemical
effects, 187
of abrasive erosion, 187
fracture (rupture) of
macromolecular chains, 187,
188, 200, 207, 214
hydrolysis, 203, 218
of fluororubber, 206
of natural rubber, 187

of nitrile rubber, 199
of polyurethane, 213
surfacial oxygenated degradation, 187, 193, 201, 210, 222
thermal decomposition (degradation), 187, 217
Surfacial mechanochemistry, 187
Systems analysis, 3
evaluation index, 4
function, 4
objective function, 4
procedure, 4
structure, 3, 35

Theory of abrasive erosion, 156
theoretical equation of wear rate, 156
prediction of wear rate, 159
energy approach, 159
fracture mechanics approach, 163
in annular pipes, 165
Theory of metal erosion
deformation-cutting, 152
fatigue, 154
microcutting, 150
Theory of rubber abrasion 85
by a line contact, 85
energy theory, 101
fatigue-fracture theory, 85
by a point contact, 42
Tribochemistry, 186
Tribology, 2
definition, 2
of elastomers, 2
definition, 2
significance, 2

Visco-elastohydrodynamic (VEHD) lubrication, 249
Volumetric wear rate, 36, 60
of erosion, 135

Wear, 33
classification, 37
constant, 37
see also coefficient of wear
definition, 33
essential characters, 34
chaos stage, 35
damage (accidental wear) stage, 35, 97
instability stage, 35
normal (steady-state) wear stage, 35, 97
running-in (unsteady state) wear stage, 34, 96
self-organization stage, 35
of metal by rubber, 227
under boundary lubrication condition, 236
wear behaviors, 227
wear mechanisms, 232
speed, 37
see also wearing intensity
Wearability, 36
see also coefficient of abrasive resistance
Wearing intensity, 37
see also wear speed
Wet (hydro-) abrasion, 39, 69
line contact, 70
multiple-contact, 72
acted by free particles (three-body), 72
acted by fixed particles (two-body), 80
point contact, 69